NATURE AND PURPOSE

John F. Haught

UNIVERSITY
PRESS OF
AMERICA

Copyright © 1980 by
University Press of America, Inc.™
P.O. Box 19101, Washington, D.C. 20036

All rights reserved

Printed in the United States of America

Library of Congress Cataloging in Publication Data

Haught, John F
  Nature and purpose.

  Includes bibliographical references.
  1. Philosophy of nature. 2. Teleology.
3. Whitehead, Alfred North, 1861-1947. I. Title.
BD581.H37   113   80-20761
ISBN 0-8191-1257-7
ISBN 0-8191-1258-5 (pbk.)

NATURE AND PURPOSE

Table of Contents

		Page
Introduction		1
I	Dualism	9
II	Physical Reality	15
III	Perception	27
IV	Emergence	41
V	Purpose	59
VI	Perishing	75
VII	Adventure	89
	Conclusion	113
	Footnotes	117

INTRODUCTION

This book takes up once again the question of nature and purpose. Realizing the controversies that inevitably accompany it, I think, nonetheless, that it is of supreme importance to raise the question again today. In the continuous outpouring of meaning into the construction of our social worlds we must ask over and over: do we have the "backing of the universe"? Or is the pursuit of meaning a striving that has no cooperation at the subhuman levels of nature? Do we carry out our projects on a stage that is blind, neutral and indifferent?

The issue is important if for no other reason than that it has a bearing on the "legitimacy" of our cultural and social worlds which are always built on the premise that purpose is worth seeking out. Our question is also important if it is true that each of us is motivated somehow, to some degree, by a will to meaning. Is this urge to find meaning that seems to be a psychological necessity really a futile stab in the dark? Or is it the welling up in human form of forces that go deep down into the rest of nature? Is there perhaps some specifiable continuity between our own felt need for purpose and whatever energies have given structure to the natural world around us and prior to us?

Further, the question of purpose in nature is of grounding significance for ethics. Issues that press today concerning the environment and the value of life cannot be separated from a fundamental consideration of the options of cosmic pessimism and optimism. Whether we can genuinely trust nature is a question underlying almost every major ethical decision we are called upon to make in our contemporary world. So, if it is possible for us to conceive of purpose in the universe in a reasonable and unsentimental manner, then by all means we should do so. Our moral instincts wither whenever they are cut off from a sense of having any roots in the cosmos.

It is hardly possible for science, all by itself, to answer our question of nature and purpose. Nor does it seem that any purely theoretical response would be convincing anyway. Our concern about purpose arises from depths of consciousness that live more comfortably with symbols and myths than with scientific or philosophical theories. And we could not hope to find purpose in nature simply by being methodologically detached and dispassionate. On the other hand, what we might be able to evince is at least the congruity between our myths of meaning and the fabric of nature disclosed by modern science and consistent cosmological theory. It would be excessively brash to state specifically what nature's purpose might be. Such a statement could arise only from a perspective we do not have. But it does lie within our capacity at least to challenge the dogmas of scientific materialism that rule out any point of contact between our myths of hope and the apparently unsympathetic world of nature that is often presented to us as the necessary consequent of a scientific approach to reality.

Natural science seems to have produced massive "evidence" to support the widespread, academically endorsed, conviction that nature is impersonal and purposeless. The following quotations from three well-known biologists represent the consensus of at least a significant number of scientific thinkers. G. G. Simpson, for example, in his book, The Meaning of Evolution, states:

> Man is the result of a purposeless and natural process that did not have him in mind. He was not planned.

. . .

Man plans and has purposes. Plan, purpose, goal, all absent in evolution to this point, enter with the coming of man and are inherent in the new evolution which is confined to him.

. . .

Discovery that the universe apart from man or before his coming lacks and lacked any purpose or plan has the inevitable corollary that the workings of the universe cannot provide any automatic, universal, eternal, or absolute ethical criteria of right and wrong.[1]

Perhaps the most extreme apology for an impersonal universe is that of the Nobel laureate, French biochemist, Jacques Monod:

> . . . chance <u>alone</u> is at the source of every innovation, of all creation in the biosphere. Pure chance, absolutely free but blind, at the very root of the stupendous edifice of evolution: this central concept of modern biology is no longer one among other possible or even conceivable hypotheses. It is today the <u>sole</u> conceivable hypothesis, the only one that squares with observed and tested fact. And nothing warrants the supposition--or the hope--that on this score our position is likely ever to be revised.[2]

And finally, S. E. Luria, in a recently popular work on biology, succinctly concurs: "The essence of biology is evolution, and the essence of evolution is the absence of motive and purpose."[3]

A number of elements in modern thought have coalesced to bring about this picture of an evolving universe impermeable to any divine purposive influence. Based especially on mechanistically evolutionist philosophies of nature, this cosmography has left us with the impression that the natural world is radically impersonal, indifferent, insensitive, blind and aimless. Even when this representation of nature is qualified by modern physics (the physics of Einstein, Bohr, Planck, Heisenberg, etc.), we are often given the same caricature of our world as indifferent, neutral or even hostile toward our deepest hopes and ideals.[4]

In the past century theologians have at times attempted to escape the embarrassment of being unable to show how God acts in the natural world as understood by science. They have tried to relate divine influence only to "history" or to inner, personal transformation. The sphere of God's action, according to a major school of Christian theology, is human "subjectivity," where in the hiddenness of free personal decision the power of God is present. But the realm of freedom and subjectivity according to this interpretation is altogether distinct from that of nature. And the possibility of God's influencing nature is off limits to theological speculation.[5]

Thus, to a great extent, theology gives the appearance of having evaded the attempt to relate nature to any divine purposiveness. It seems to have overlooked the obvious fact that every historical event or inner, hidden act of decision and conversion is simultaneously an act occurring within the web of events that make up what science calls "nature." Harvard theologian Gordon Kaufmann is emphatic on this point:

> It is impossible to speak of history as though it were a realm of freedom and decision entirely separate from nature. Certainly the biblical perspective is not characterized by such nonsense. It is a measure of the desperation of contemporary theology and faith, in the face of the power of the modern scientific world view . . . that this way out was attempted at all.[6]

And yet almost every time theological speculation does give birth to hypotheses as to how God influences nature, a host of troublesome questions arises. Why, for example, is the natural world so abundantly inhabited by elements of chance, waste, evolutionary catastrophes, dead ends, struggle for survival, indeterminacy, and other baffling phenomena? If God acts in the natural world, how is it that things have gotten so far out of control? Or does God act only occasionally, locally and temporarily? If so, then is not such a deity capricious, incapable of inspiring worship or gratitude? Further, how can a spiritually transcendent being influence material reality? What kind of causation would be operative in that case? If God is causally related to nature, why is there no overwhelming evidence of it? If God acts creatively, as reported in biblical religion, why does the world come into being over a period of billions of years in an evolutionary way, with no obvious directionality to the process? These and other questions immediately confront any attempts to render intelligible the religious symbols of divine creative or redemptive activity in nature.

A general crisis of meaning in the intellectual world has accompanied the emergence of the picture of an indifferent universe, barren of all purpose. And a great many strains of modern art, philosophy and literature have blossomed in the past two centuries in order to justify human life in this alien world. Some of these expressions have appealed rhapsodically to the ancient tragic vision of existence for support against the indifferent universe. Bertrand Russell provides one of the more familiar of these dismal reactions. Against the hostile world, a world that was not made for us and is unworthy of us, the individual soul, he says, ". . .must struggle alone, with what of courage it can command, against the whole weight of a universe that cares nothing for its hopes and fears."[7]

Underneath of the extravagant cosmic pessimism of Russell and those of like mind there is a persistent acquiescence to the conventional materialistic view of the universe said to be rooted in scientific method. This view has been held to be unquestionably sound since it has been arrived at by the generally accepted procedures of scientific inquiry. And challenges to it are usually ignored as reactionary naivete.

In this book I hope to present a respectable challenge to the orthodox cosmography that underlies current views (such as those of Russell, Simpson, Monod, and Luria) that nature is inherently recalcitrant to purpose of any sort. My argument will develop initially by way of criticizing the dualism that constitutes the mythic substrate of this academically enlightened conviction. I am aware of course that much modern philosophy, theology and literature has already taken on an anti-dualistic cast, and I am in sympathy with it. However, usually the protest against the Cartesian isolation of subjectivity from nature is excessively romantic and intellectually feeble.[8] This protest is based on sound feelings that dualisms are unstable and unsatisfying. But the manner in which it argues for a natural world that shows signs of divine care is often philosophically unconvincing.[9] I hope that in the following pages a consistent philosophical position, compatible with common sense, science and sound logic will emerge as an alternative to dualism and its offspring, scientific materialism.

In developing my position I shall utilize insights of a number of respected thinkers. Foremost among these will be Alfred North Whitehead, renowned mathematician and scientifically sophisticated philosopher, who developed an elaborate cosmology and metaphysics in protest against dualism and mechanism. I shall also call upon my reading of Michael Polanyi, Bernard Lonergan, Charles Hartshorne, John Cobb and many others for support in my presentation. It will not be possible for me always to isolate clearly the contributions each of these has made to the following discussion. For I shall be synthesizing their thought with my own and with the relatively simplified vocabulary to be employed in the argument. I hope that I do not flagrantly reduce or dilute their thought in presenting my position to a more general readership than their own works often allow.

Especially in the case of Whitehead's thought some over-simplification is essential if readers are even to begin an appreciation of his work. Thus I have drastically modified his terminology, preserving intact only a few important expressions and axioms, while still utilizing many of his ideas.

Finally, while I am impressed by the philosophy of nature and a great many of the insights into religion that Whitehead offers, I am not able, so far as I understand them, to align my thought with every aspect of his very undeveloped reflections on God. I accept these reflections, however, as suggestive rather than definitive and regard them as deserving of further development and criticism. Throughout this work the reader will note that I am indebted to Whitehead's descriptions of physical reality, perception and causation. In presenting my own reflections on God's relation to nature, however, there are several points where I shall have to depart from Whitehead's approach.

Chapter I

DUALISM

Dualism is a way of thinking about reality and man that separates each of these into mutually exclusive spheres. Dualism divides reality into spirit and matter and man into soul and body or into mind over against matter. A tendency toward dualism in feelings, in mythic-symbolic consciousness and in philosophical reflection has always been present in Western culture (and perhaps other cultures also). It is this tendency more than any other single factor that lies behind the conviction of scientism that nature is indifferent to human hope and to divine influence. In its doctrines of a purposeless universe running blindly toward final catastrophe, scientific materialism unknowingly exhibits the atavistic inclination toward dualism that is a perennial ingredient of our consciousness. This dualistic predilection, therefore, merits a brief examination at the very beginning of our inquiry into nature and purpose.

Dualism and the Problem of Life

There is an interesting and ironic history behind the modern conviction that physical reality is fundamentally dead or inert, and, therefore, blind, aimless and impersonal. The persuasion that nature is purposeless rests on the premise that, prior to man's evolutionary appearance, nature is totally lacking in anything like mentality. The universe is "unconscious" according to this conviction in all of the evolutionary phases preceding the emergence of the noosphere.[1]

It is important that we grasp clearly how the picture of an "unconscious" universe" evolved out of the myths and philosophies that wrested mentality from its matrix in nature. The key to our story lies in the inveterate dualism of Western thought and in the inversions and monistic resolutions that this dualism has undergone.

Early man's experience was saturated with a sense of the pervasive aliveness of the world. Very little that resembled the inert figured into the horizon of his consciousness. People, plants, animals, rivers, weather and other natural conditions overwhelmed him with an impression of the vital fluidity of it all. "Primitive panvitalism," Hans Jonas writes, "was the comprehensive view."[2] But, if everything was alive, then the main occasion for wonder and primitive bafflement was the fact of death. If everything lives, how can this or that dead body appear to be so inert, so lifeless?

> This is the paradox: precisely the importance of the tombs in the beginnings of mankind, the power of the death motif in the beginnings of human thought, testify to the greater power of the universal life motif as their sustaining ground: being was intelligible only as living; and the divined constancy of being could be understood only as the constancy of life, even beyond death and in defiance of its apparent verdict.[3]

In archaic thought, and even up to the Renaissance, death was considered somehow illusory, incapable of being made intelligible in terms of the overwhelming flux of vitality that buoyed human existence. But this was before Copernicus, Galileo, Newton and Descartes, all of whom contributed, along of course with many others, to the birth of the modern picture of the overwhelming inertness of matter. Modern physics, geology and astronomy have disclosed vast tracts of empty and lifeless space. And they have led us to conjecture how precious and precarious is the infinitesimal amount of life that exists in the universe.

Modern thought, as a result, has a radically different perspective from that of primitive panvitalism:

> Death is the natural thing, life the problem. From the physical sciences there spread over the conception of all existence an ontology whose model entity is pure matter, stripped of all features of life. What at the animistic stage was not even discovered has in the meantime conquered the vision of reality, entirely ousting its counterpart. The tremendously enlarged universe of modern cosmology is conceived as a field of inanimate masses and forces which operate according to the laws of inertia and of quantitative distribution in space. This denuded substratum of all reality could only be arrived at through a progressive expurgation of vital features from the physical record and through strict abstention from projecting into its image our own felt aliveness.
>
> . . .
>
> This means that the lifeless has become the knowable par excellence and is for that reason also considered the true and only foundation of reality. It is the "natural" as well as the original state of things. Not only in terms of relative quantity but also in terms of ontological genuineness, nonlife is the rule, life the puzzling exception in physical existence.[4]

As a result of this inverted theoretical situation, Jonas concludes" . . . it is the existence of life within a mechanical universe which now calls for an explanation, and explanation has to be in terms of the lifeless."[5]

The explanation of the "living" in terms of the nonliving has become the ideal of much modern scientific inquiry. It is axiomatic to many, for example, that biology is reducible to physics and chemistry, and therefore, that life is reducible to the "inanimate." The dizzying advances in molecular biology blur the former distinctions between man, animal, plant and mineral; and the recent "reductions" of mind to brain are fruits of the methodological imperative to explain the animate and mental in terms of the inanimate and the unconscious.

It is the dualism of soul and body, spirit and nature, mind and matter that has made possible the shift of problematics from that of how to explain death if everything is alive, to that of how to explain life if everything is dead. Dualism is the pivotal mythic and philosophic construct on which this inversion has turned.[6] While dualism has been an important factor in our coming to vivid awareness of the faculty of mentality which makes us distinct and aware of our special status, it has at the same time exorcized nature outside of human mentation of the qualities of mind and aliveness that we experience in the subtlety of our own consciousness. It has given rise to an "ontology of death."[7] Anything that is not "mind" "in here" is denuded of the vitality associated with thought and experience, and is placed "out there" in a totally different kind of world of inert, passive material objects.

This bifurcation of reality came to expression in Orphism, Gnosticism, Neo-Platonism, and Medieval anthropology, culminating in Descartes' noted distinction of <u>res cogitans</u> (mind) and <u>res extensa</u> (matter). The influence of the dualistic myth and metaphysics on the birth and growth of modern science has been amply documented, and it is not necessary to trace the whole story here.[8] It is enough only to point out that the inertness bequeathed to matter by dualism has become the basis upon which the mathematization of physical reality and motion in Newtonian and Cartesian physics has been constructed. And it is safe to say that without the dualistic premise modern science as we know it could not have developed as fruitfully or rapidly as it has.

It is also true, however, that dualism still lurks behind the dominant contemporary philosophies of nature in which matter remains essentially mindless and lifeless. It is dualism that, in the final analysis, structures the current methodological ideal in the life sciences which consists of the attempt to specify or explain biotic and conscious operations in terms of the sciences (physics and chemistry) that deal with the allegedly inanimate. Without the sphere of unconscious and lifeless chunks of matter delineated by dualism such a methodological ideal (which animates current efforts especially in biology to find the physico-chemical "secret" of life) could hardly have taken hold in scientific inquiry. In a curious way we owe a great deal to what I think is a serious mistake in cosmology.

Dualism and the Problem of Purpose

Standing at the end of the history of this dualism it is easy for us to see why any attribution of "mentality" (and therefore of purposefulness) to nature will be dismissed as romantic anthropomorphism. By expelling anything that resembles feeling, experience or perceptivity from the sphere commonly called nature, modern thought has also eliminated the possibility of attributing purpose to nature also. It has rightly recognized that without a vein of "mentality" in the universe there can be no purpose either. And so, by rejecting the alliance of nature and mind, it has removed to that extent the feasibility of our searching for purpose in the world of nature. For where there is no dimension of "mind" there is no aim or purpose either.

In turn, the rejection of a teleological

universe has led in many cases to doubt about any human purpose whatsoever. Needless to say, there has been a close connection all along between the modern experience of meaninglessness and the development of the picture of an impersonal universe that gives no backing to our projects. The same dualistic myths that have made us feel exceptional have also led to our sense of alienation from nature and purpose.

It is possible in theory to anticipate, therefore, the enormous implications that a new alliance of nature and mind might have for the contemporary crisis of meaning. Nothing less imposing than the significance of our lives is bound up with the quest for a union of mind and nature established on solid grounds compatible with reason, common sense and science. If we could grasp somehow that our subjectivity is a blossoming forth of nature itself, and not some enigmatic "nothingness" or separate substance over against nature, we would have at least the context in which to discuss once again the question of nature and purpose.

The basis for a synthetic vision of mind and nature is worked out most comprehensively by Alfred North Whitehead. The following chapter is an introductory synopsis and simplification of his notion of physical reality. It is designed especially to highlight his critique of dualism and to prepare for our subsequent discussion of the central problem of science and religion today, that of purpose in nature.

Chapter II

PHYSICAL REALITY

The question, "Is there purpose in nature?" must be preceded by another: What is meant by "nature"? R. G. Collingwood, in his well-known book, The Idea of Nature, develops the thesis that the idea of nature in philosophical discussion has always been conditioned by historical preoccupations and circumstances.[1] We cannot hope to isolate nature from our historicity so as to describe clearly and distinctly what it is "in itself." Even the statement that "nature is objective" and presumably neutral toward human meanings, is the product of an historically rooted perspective, the very one that we called dualistic in the previous chapter. The idea of nature, then, is always already overlaid with our meanings and cannot be understood apart from them.

Instead of constituting a barrier to our understanding, this interpenetration of our meanings and mentality with nature is our very access to understanding physical reality. Whitehead observes that "scientific reasoning is completely dominated by the presupposition that mental functionings are not properly part of nature." But this reasoning, he says, while justifiable within limits, leaves out a massive amount of data:

> . . . this sharp division between mentality and nature has no ground in our fundamental observation. We find ourselves living within nature. ... We should conceive mental operations as among the factors which make up the constitution of nature.[2]

Throughout this book we shall follow

Whitehead's admonition to avoid the dualism that puts nature in one arena and subjective experience in another. There is no decisive line of demarcation in the universe that segregates experience on the one side from insensitive objects on the other. Rather, as we shall propose, the universe is ultimately and pervasively made up of units of experience.

The alliance of nature and mind for which we are arguing is presumed in an ambiguous way by science itself. Evolutionary theory, physiology, molecular biology and other fields of science have made our species intelligible to us only as a part of the natural world. One of the major implications of this naturalistic absorption of the human dimension is that the categories of thought we employ to understand nature are not irrelevant to our understanding the nature of man. And a great deal of contemporary anthropology, sociology and psychology utilizes this premise. At the same time, however, the obverse should also hold: the concepts we employ in order to interpret human experience are applicable in our attempts at understanding the rest of nature. This latter suggestion, however, has not often been taken as seriously as the first. But if we are to understand anything about nature we must not exclude ourselves and our experience from the world we are trying to understand.

Suppose, then, that we begin our description of the natural world by making explicit what it means to be an experiencing, conscious subject. Is this an unscientific way of proceeding? Perhaps so, depending upon one's view of what science is. Nevertheless, it seems to be a very _empirical_ way of beginning to understand the universe, more radically empirical in fact than science is itself.[3] Here we are attending not only to the data of sense-perception but also to a much more proximate set of givens--the experiential components of our own subjectivity. These data constitute the nearest and richest point of entry each of us has to the physical world in which we are embedded. Our attempt to grasp what nature is will be considerably emaciated if we abstract from these data. The fact that awareness of our experiencing subjectivity is so immediate to us is no warrant for our shoving it aside when we work out our cosmological perspectives. After all, our own experiencing, knowing and desiring are a part of the world-process, and we cannot hope to understand this process without first reflecting upon them as somehow expressive of what the world is.

The Subjective Experience of Becoming

One of the most immediately available characteristics of my subjective experience is its quality of "becoming". This becoming is so basic and pervasive that I seldom reflect upon it. But retrieving it now on a reflective level will give us the needed base for understanding our physical universe.

Each moment of my life is made of becoming and perishinng. As I reflect on the experience of my temporality I note that it is a process composed of occasions, each of them a "throb of experience"[4] that "momentarily" becomes and then perishes. My own enduring subjectivity is made up of these moments of experience, each flowing into the next as part of an enduring series of experiential occasions.

While I always abide in the present, the present moment is always perishing. As it perishes, however, it is not totally thrust into nothingness. Something of my experience of past moments abides in the "now". I am able to retain in memory that which no longer is now, the moments of my experience that have become and then perished. And each moment of my experience somehow preserves, however vaguely, the moments that have made up my experience in the past. At the same time that I hold onto the past in memory, I also anticipate the future that is not yet. I grasp yet unrealized possibilities, and my present feelings and decisions take a shape corresponding to my accepting or refusing these possibilities.

There is nothing extraordinary in this brief description of temporal experience. And yet there is the nucleus here of a model we might employ to understand the basic constituent elements of physical reality. On the basis of our own experience of temporality we may construct a general view of nature that will allow us to reconsider this issue of purpose in the universe.

What Makes Up the Universe?[5]

Once we reject the dualism of mind and nature we must place our experience of temporality on a continuum with the rest of physical reality. What we ourselves experience as moments of duration must bear some relation to the rest of nature. Therefore we may entertain the hypothesis that the universe as a whole is made up of moments of experience analogous to those that constitute our own enduring and becoming. Following Whitehead's terminology we shall refer to these experiences, as "events" or "actual occasions" of which our own experiences are only one variety.[6] It is these events or occasions of experience that make up the universe of nature. And each of these units of becoming is characterized in its own way by the features that make up the occasions or moments of our own experience. Something like present immediacy of enjoyment, memory and anticipation enters into each of the events that organically compose physical reality.

Initially the suggestion that the universe is composed of experiential events will understandably seem quite puzzling. Our common sense tells us that the universe is made up of chunks of solid matter simply located in space and time. Furthermore, the classical physics of Galileo, Descartes and Newton, basing itself on this common sense view of matter, portrays nature as made up of hard, impermeable material particles or mechanisms obeying immutable physical laws. This "corpuscular" view of nature is firmly fixed in the minds and sensitivities of most of us, including biologists, physicists and chemists. So when it is maintained that experiential temporal events, rather than spatialized particles, are the ultimate constituent elements of the universe, our first reaction will probably be somewhat skeptical.

However, let us examine the traditional philosophy of nature based on classical physics and common sense. Upon closer inspection this corpuscular, spatialized view of nature that sticks so firmly in our imaginations turns out to be incompatible with experience, contemporary science and sound logic:

(1) The Testimony of Experience: The persistent belief that the physical universe is unexperiential can be sustained only so long as we arbitrarily and dogmatically exclude from nature our own experiencing subjectivity. The radical empiricism alluded to above demands that we attend to all the data of our experience including the most proximate, our experiencing itself. Now in our experience of our own subjectivity we do not discover anything like the inert brute stuff into which classical physics attempts to analyze nature. Scientific materialism either abstracts from this phenomenon of subjective consciousness, or else, when it does advert to it, attempts to explain it only in terms of what it takes to be inanimate matter. Thereby it is led to the dubious view that the unconscious is the cause of the conscious. The procedure we shall follow, however, is one that begins with our own mental experience as the key to the rest of nature. We object to the materialistic approach as unempirical and excessively abstract in its representation of physical reality. Scientific objectivism, it appears, pays only lip-service to the empirical imperative. It brackets as useless to science what is the most lively and immediate experience we have of the world--our own feeling, becoming, conscious and desiring subjectivity. Thus it fails to reach a fundamental description of the composition of nature.

(2) **The Implications of Contemporary Science:** If the conclusions of modern biology and physics were fully thought out it would be extremely difficult to reconcile them with the classical materialistic philosophies of nature. It appears, though, that orthodox biology has not yet torn itself away from the old cosmography, nor for that matter has modern physics succeeded totally in doing so either. A preference for atomicity (explanation in terms of amino and nucleic acids) prescribes the method for molecular biology; and a persistent materialism hovers over the essentially anti-mechanistic physics of this century. The common sense notions presupposed by the scientists of the seventeenth and eighteenth centuries, and that still shape our everyday thinking, are no longer tenable as comprehensive and fundamental explanations of things. And yet they continue to infiltrate the many worlds of thought. What Whitehead wrote four decades ago regrettably still holds today:

> The common sense notion still reigns supreme in the work-a-day life of mankind. ... The state of modern thought is that every single item in this general doctrine is denied, but that the general conclusions from the doctrine as a whole are tenaciously retained. The result is a complete muddle in scientific thought, in philosophical cosmology, and in epistemology. But any doctrine which does not implicitly presuppose this point of view is assailed as unintelligible.[7]

Issues in science and religion today arise in great measure from this general intellectual muddle. The problem of purpose in nature in particular stems out of the reluctance of contemporary thought to revise the Newtonian-Cartesian cosmography in the light of recent developments in physics. Whitehead noted that "... in the present-day reconstruction of physics fragments of the Newtonian concepts are stubbornly retained. The result is to reduce modern physics to a sort of mystic chant over an unintelligible universe."[8] But if we were to take more seriously the modern revolution in physics we would have to move radically beyond the corpuscular philosophy and the atomistic ideal that seek to divest nature of mentality and significance.

In its notions of matter, space and time, the new physics gives us a fundamentally different picture of nature from the one we are accustomed to. For example, matter is now seen as the product of a "group of agitations" or "energy-events" rather than a collection of bits of solid stuff.[9] And because matter emerges out of a substratum of energetic occurrences, we also have to revise our notions of the relation of both space and time to physical reality.

As far as the question of space is concerned, in the universe of classical physics

> ...the concept of matter presupposed simple location. Each bit of matter was self-contained, localized in a region with a passive, static network of spatial relations, entwined in a uniform relational system from infinity to infinity and from eternity to eternity. But in the modern concept the group of agitations which we term matter is fused into its environment. There is no possibility of a detached, self-contained local existence. The environment enters into the nature of each thing. ... In truth, the notion of the self-contained particle of matter, self-sufficient within its local habitation, is an abstraction.[10]

Accordingly, "any local agitation shakes the whole universe."[11] As the poet, Francis Thompson, put it:

> All things by immortal power
> Near and far
> Hiddenly
> To each other linked are,
> That thou canst not stir a flower
> Without troubling of a star.[12]

The testimony of the poet is no longer alien to the scientific view of the world.

The dramatic revisions in our ideas on the spatiality of matter have, therefore, a major implication: We must abandon the assumption that we can understand physical reality by simply locating bits of matter in space without taking into account the relational web of energy-events in which they are situated. The "assumption of simple location," upon which classical physics was based, abstracts from an aspect of physical reality that must now be considered fundamental and not just accidental--time.

"There is no nature at an instant."[13] In order for matter to take on the shape that it does, time is required. Time is woven into the texture of things in a much more interior way than the old cosmology allowed for. According to the latter, if time abruptly came to a halt, there would still be a universe of "space" filled with bits of matter. That is, there would be nature at an instant. But this sort of universe is inconceivable according to the space-time physics of today. It is commonly recognized now that the things we call molecules, atoms, electrons, neutrons, protons, etc. are all physical *patterns* that require time in order to make their impact felt. Without a certain quantum of time an electron simply could not be. Electrons, and all other so-called sub-atomic particles, are the result of happenings. And these happenings are spatio-*temporal* vibrations that cannot be simply located. Nor can they be adequately described at all without taking into account their total context--the temporal, evolutionary universe.

Thus contemporary physics supports the proposal made earlier that events rather than solid particles of inert matter are the fundamental units of nature. Particles, as it now appears, are really abstractions useful for grouping certain patterns of occurrences. But it is the energy-events themselves that are the actual fundamental units. This conclusion calls for a whole new way of picturing and representing the world of nature. Unfortunately, though, modern scientific thought has barely begun the difficult but essential enterprise of revising our cosmography. To a great extent it continues to fall back on the classical imagery even while going far beyond it theoretically and mathematically. Some re-imaging, however, is a pre-requisite of any possible vision of purpose in the universe.

(3) <u>The Demands of Logic</u>: The materialistic philosophy of nature fails to take into account not only our experience of subjective consciousness and the modern revolution in physics. It also fails on logical grounds by confusing the abstract with the concrete. That is, it is guilty of what Whitehead calls the "fallacy of misplaced concreteness."[14] Following the assumption of simple location,[15] the cosmology derived from Galileo, Newton and Descartes persistently views the objects isolated by scientific method as though they were the fundamental units of the physical world itself. In fact, however, they are scientific abstractions derived by distinguishing "primary qualities" (mass, position, velocity) of large bodies from "secondary qualities" (color, taste, sound, smell, etc.). Primary qualities are viewed as objective, i.e., independent of the knower's frame of reference, while secondary qualities are judged to be subjective, i.e., involving the complicity of the subject imposing his own peculiar sensory apparatus on the bodies perceived. Along with this rigorous distinction of primary from secondary qualities there has often gone a belief that only the primary qualities could be called really real (their persistence throughout accidental changes being the

criterion of their reality), and that secondary qualities are not part of the real world.

This division is obviously rooted in the dualistic expulsion of experience from nature. It shows vividly the reluctance of traditional reflection to accept our own perceptivity as itself part of nature. And in doing so it leads to a serious error in logic: after abstracting so completely from the experiential quality that pervades all of nature it sets forth the desiccated end-product of its abstracting as though it were reality-itself and everything else a mere coloring by human sensory projection.

This misrepresentation is a result of the fallacy of misplaced concreteness, the confusion of the concrete with the abstract. This is a fallacy to which we are always inclined by our talent for reductionism, but one that has wrought a pernicious influence in modern thought. For example, it has encouraged the uncritical attempt to explain experience, life and consciousness exhaustively in terms of colorless and inanimate abstractions. In order to grasp more vividly the way in which the fallacy of mistaking abstractions for concrete realities has inclined thought in this direction, observe the following "rough" breakdown of nature's hierarchical structure:[16] i.e.

(1) human life
(2) animal life
(3) vegetable life
(4) single living cells
(5) large scale inorganic aggregates of occasions
(6) energy-events disclosed by modern physics

The first five of these types of natural occurrence are easily accessible to sense perception (aided perhaps by instruments of observation); the sixth, however, is not available to ordinary sense perception. The events, or actual occasions (6) that we have located at the ground level of natural occurrence are in an entirely different order from the aggregates (spatially located) that constitute the fifth category. And yet it is the latter (5) that have usually been employed by scientists in their attempts to understand categories 1-4 in a fundamental way. Unobservable but dynamic physical events are in fact patterned in inorganic aggregates in such a way that these aggregates are inert, lifeless, unconscious and aimless. But then the visible bodies (5) are imaginatively decomposed by science into smaller, invisible particles possessing the same obdurate features as the larger aggregates on which they are modeled. Thus, when the universe is "explained" in terms of these peculiar contrivances the apparent aimlessness and lack of mentality that we read out of rocks and grains of sand is read back into the universe as a whole and into its constituent occasions.

The fallacy in this projection consists of confusing the abstracted features of aggregates with the concreteness of individual occasions of experience that make up the aggregates. The objects of our ordinary experience, things such as rocks, trees, animals and persons are composites or groupings of what we have been calling occasions of experience. In various modes of serial ordering, the world's constituent occasions experience one another so as to form these various assemblages of occasions. Our sense perception, however, refers us only to inorganic aggregates or to living and conscious "societies" of these occasions. It is incapable of breaking them down into the experiential-mental moments that lie beneath the threshold of what we can perceive with our senses. Consequently, our notion of physical reality suffers from our taking sense perception too one-sidedly as the foundation of cosmological speculation. The consequences of this superficial notion of perception will be set forth in the next chapter.

Conclusion

If nature is in any sense purposive then "mentality" (not necessarily consciousness) would have to be a pervasive and not merely a localized, accidental and fragmentary characteristic of it. By mentality, in this broad sense of the term, is signified the quality of active receptivity to meaning, value and significance that is intrinsic to each experiential occasion. In order to allow for this quality in nature we must be prepared to envision its constituent elements as themselves units of perception or "feeling", that is, as having rudiments of mentality as we know it from our own experience. In this chapter we have proposed a notion of physical reality in which mentality is intrinsic to the physical. In such a conception the natural world is an organismic one where the occasions that make it up are bound together in mutual, internal relatedness by virtue of their capacity for experiencing (prehending) one another. Rendering mentality a universal category of reality sounds strange when viewed from the perspective of our dualistic heritage and from that of the conventional materialistic view of physical reality. And as we shall see, it also seems alien to the typical notion of perception that accompanies the classical cosmography. Perhaps, though, some of this foreignness may be removed by a careful examination of the nature of perception.

Chapter III

PERCEPTION

It should be evident by now that our question about purpose in nature is one way of raising the problem of the intelligibility and validity of religious discourse in a scientific age. Is the religious reference to an ultimate ground of meaning compatible with what science tells us about our world?

Religious articulation of meaning has often been associated with "teleology". Derived from the Greek word "telos" (goal, end, purpose), the term "teleology" refers to the branch of cosmology that treats of the purpose or destiny of the universe. Biblical religion in particular, because of its eschatological and at times apocalyptic characteristics, has often been viewed as teleological; and in a qualified sense this association is appropriate. Precisely how to understand and pull together the wide variety of ways in which the Bible symbolizes the purposes of peoples, nature and history has always been a problem. But it is safe to hold, at least in a loose sense, that biblical consciousness is teleological in that it posits symbolically a final significance to events in nature and history.

Important devotees of modern science, however, have found it hard to accept the believer's apparently unwarranted espousal of such a teleological perspective on the universe. Scientific method, they maintain, discloses no evidence of cosmic purpose. While it is evident to science that there is a functional "teleonomy" or machine-like purposiveness in individual organisms (for example, the fish's eye is constructed so as to enable it to see under water, the heart toward pumping blood, the human brain toward problem-solving, etc.), still there is no hard evidence that life itself, terrestrial evolution or the universe as a whole has any overarching meaning. Belief in a teleological universe is viewed as wishful thinking rather than verified knowledge. As Jacques Monod would have us understand it, teleology is simply the product of projecting onto an intrinsically indifferent universe our own intensely teleonomic nervous systems.[1] According to Monod's interpretation, those of us who are inclined toward teleology extend our intimate perception of the limited teleonomy in our own organisms onto the foreign terrain of the "objective" universe. And out of such groundless projections are fashioned the myths, religions and philosophies that have for ages given people a false sense of warmth and purpose in the cosmos. Scientific method, though, cannot reconcile itself to teleological perspectives and, therefore, must reject any such facile covenants of man with a world that is alien to his longings for ultimate meaning.[2] Monod's position is reminiscent of innumerable others that see the reading of purpose into nature as analogous to our subjectively superimposing colorful secondary qualities onto starkly colorless "objective" and neutral primary qualities. Accordingly teleology is characterized as essentially derivative, subjective and flawed with the arbitrariness and relativity that pertain to secondary qualities. The view that nature is purposeful appears to have no grounding in the universe itself.

I would like to explore further the presuppositions of this common "scientific" objection to teleology. While I would admit that scientific materialists are correct in their rejection of certain rigid forms of teleology, I shall argue here that their opposition to a universe imbued with value and significance flows in part out of a naive notion of human perception. This notion of perception, in turn, is linked to the dualism discussed in Chapter I and the materialistic view of physical reality criticized in the preceding chapter.

Is Sense-Perception Primary?

Michael Polanyi has accurately and tidily stated the crux of the problem of science and religion:

> Intellectual assent to the reduction of the world to its atomic elements acting blindly in terms of equilibrations of forces, an assent that has gradually come to prevail since the birth of modern science, has made any sort of teleological view of the cosmos seem unscientific and wool gathering to us. And it is this assent, more than any other one intellectual factor, that has set science and religion... in opposition to each other in the contemporary mind.[3]

Polanyi's statement brings to mind the similar one by W.T. Stace issued several decades ago:

> Religion could survive the discoveries that the sun, not the earth is the center; that men are descended from simian ancestors; that the earth is hundreds of millions of years old. These discoveries may render out of date some of the details of older theological dogmas, may force their restatement in new intellectual frameworks. But they do not touch the essence of the religious vision itself, which is the faith that there is plan and purpose in the world, that the world is a moral order, that in the end all things are for the best. This faith may express itself through many intellectual dogmas, those of Christianity, of Hinduism, of Islam. All and any

of these intellectual dogmas may be destroyed without destroying the essential religious spirit. But that spirit cannot survive destruction of belief in a plan and purpose of the world, for that is the very heart of it. Religion cannot get on with a purposeless and meaningless universe.[4]

The intellectual assent to an indifferent and aimless cosmos is usually defended on the assumption that our senses give us such a world and that in all honesty we should accept this world in the precise shape our sense perception transmits it to our minds. We are often exhorted by scientists and philosophers alike to accept the material given to us by sense perception as though it is the rock-bottom foundation of our knowledge of the physical world. Simultaneously we are told to refrain from coloring neutral sense data over with our subjective wishes and teleological desires.

Are we are obliged to accept this doctrine that sense perception is the ultimate foundation of full and genuine knowledge of the world? It seems so obvious to most of us that our knowledge of the world enters first through the gateway of the five senses, especially sight, that any questioning of the primacy of this kind of perception will initially seem quite bewildering. Yet it is not self-evident, after all, that sense perception is primary. And a careful investigation of our own experiencing may lead us to question the common philosophical assumption that it is. What is more, a re-experiencing of our experience without the blinders put on us by several centuries of "sensationalist" philosophy can lead us to question the legitimacy of the picture of an indifferent, totally non-teleological universe apparently entailed by this truncated rendition of human experience.

The Two Poles of Perception

When "perception" is limited to the material presented to our minds by the five senses we are by no means dealing yet in a <u>fundamental</u> way with the reality of the world. Instead we are simply focusing on the end results of a very complex process of experiencing. This process has filtered out and abstracted from the data presented to us at a more basic level of our being by a much more global mode of sensitivity. We seldom think of our perceptive experience as itself a <u>process</u> of screening and abstracting. The data presented by our five senses appear to be so clear, distinct and irreducible that we scarcely recognize the experiential refining process <u>prior to</u> sense perception. And we seldom notice that the price paid for clarity in our sense perception is the lack of vivid awareness of what we have left behind in the process. This lack of awareness, though, is directly commensurate with the "perception" of vacuousness that scientism has "found" to be "inherent" in the physical world. The indifference and intrinsic valuelessness of the universe of scientific materialism is the product of a view of perception that ignores the depths and hazy beginnings of a whole process of sensation that merely culminates in but goes far deeper than sense perception.[5]

Perception may be understood as having two poles, primary and secondary. The first pole (called "perception in the mode of causal efficacy" by Whitehead) we shall refer to simply as "primary perception". At this pole of the perceptive process there is a pervasive and vague feeling of the influence of the world upon our being and becoming. As the universe first influences the percipient subject it is not yet parcelled out into distinct objects, chunks of matter, particles or any other merely spatially defined phenomena. Instead, at the initial point of entry into primary perception the universe is felt viscerally as a value-laden causal process of events, a series of occasions of experience with which the subject's bodily sensitivity is itself continuous.

Again, this sounds extraordinary. Conditioned by dualism, we usually think that there are two sharply separated kinds of being, subject and object, mind and matter. Accordingly, we tend to believe that matter impresses itself (somehow?) on mind and that this impression is what constitutes perception. Often we go even further and identify the objects of our experience as constellations of the particles defined by science. Thus we assume the priority of material "stuff" over experience and make experience a passive reception of "matter" by our senses. But this common conviction about the nature of perception is in the final analysis incoherent. It never explains exactly how the experiencing subject actually comes into contact with the inert matter that makes up the objects of sense perception. There remains an unbridgeable gap between two totally disparate types of reality, mind and matter. How matter makes the transition into mind and how mind receives content or meaning from the mindless realm of matter is never clarified. We can avoid this epistemological obscurantism if we recognize the experiential, perceptive quality of all reality. With Whitehead we have made occasions of experience the basic constitutive elements in our envisagement of the universe. Since experience is pervasive there can be no isolation of mentality from nature. Every occasion is itself a moment of sensitive "enjoyment". Each "drop of experience" receives the entire universe into itself more or less vaguely. Although the immediate past of an occasion of experience is felt or experienced with more intensity than the remote past, still in a dim way the whole universe and its past is synthesized into each of the moments of experience that taken together make up reality.

Our own experience of the world is not an exception to but an exemplification of the features that pertain to all of nature's becoming and experiencing. Thus in what we have called primary perception there is a feeling of the entire universe entering into our experiencing. However, only a very small sector of this universe is apprehended with any degree of vividness. It is the function of the <u>secondary</u> pole of perception (in the mode of "presentational immediacy") to project the spatially clear and distinct objects of sense perception onto the background of the temporal series of occasions that are vaguely assimilated by primary perception. The so-called objects of secondary perception, therefore, are <u>abstracted</u> from and projected onto a densely dynamic field of occasions that can never be fully brought into focus but that continue to enter our perception at the primary pole.

Science takes as the material for its inductive and descriptive procedures the spatialized abstractions of secondary perception. As such, science is incapable of dealing with nature in a fundamental way. This is not to deny that scientific method is appropriate and valid. Rather it is simply to point out its insufficiency for describing the world in depth and with anything approaching adequacy. Since it describes and predicts on the basis of correlations among the relatively abstract entities of secondary perception, it cannot be taken as comprehensive or sufficiently rudimentary. While modern physics has taken us deeper into the eventful, relational spatio-temporal web of nature than did classical physics, it too is far from giving us a fundamental cosmological description. Science always abstracts to some degree from the universe and its significance as it is apprehended in primary perception.

And yet it is difficult to find many important scientific thinkers, philosophers of science or naturalists who sufficiently recognize the failure of scientific method to reach as deeply into the nature of the universe as we must if we are to respond intelligently to the question of purpose. More often than not philosophers of nature take the abstractions of science and secondary perception as the bed-rock of their speculations. And since these abstractions lack the aspects of mentality and value that we would locate at the deeper level of primary perception, the universe of science appears as essentially mindless and insignificant. Seemingly it lacks the prerequisites to sustain a teleological interpretation.

Science of late has been inundated by the disclosure of ever new and peculiar sub-atomic "particles." There is still alive an underlying hope that we will discover some ultimate particulate substance out of which we might conceive the building up of the universe into the diverse phenomena that confront our senses. In <u>cosmological</u> description, however, we cannot pretend that by coming upon some irreducible particle or pattern of particulate activity we will have reached a firm foundation for a philosophy of nature. For what we experience primordially are not particles but rather occasions of experience bound together serially into enduring objects, particles, corpuscular "societies" or personal "societies." The search by contemporary physics for more and more basic particles is a useful and exciting part of scientific work. Nonetheless, it does not considerably deepen, but merely broadens our sense of what the universe is. Cosmology demands that we go beneath the scientific abstractions called particles, including those strange and elusive ones of recent physics.

Our thesis is that experience itself is the basic "building-block" of the universe. Accordingly our primary perception is not something over-against an objective, insensate material world "out there." Rather our own perceptive subjectivity is itself a blossoming forth of the world process, totally continuous with its intrinsically eventful and experiential character. We can understand nature as itself intrinsically perceptive rather than as a body of opaque particles. And this means that our universe is quite different from the congeries of mindless abstractions that inhabit the world of scientific materialism.

We stated earlier that a mindless universe is a purposeless one. There is no way in which a hypothetically teleological principle could influence or be felt by a natural world that lacks any quality of sensitivity to such influence. However, there is no convincing reason for us now to believe that the non-teleological universe issuing from scientific method's rejection of final causal explanation coincides with the one given to us at the pole of primary perception. Our primary perception of the world is much deeper, more sweeping and more ragged at the edges than the lucid sensing by our eyes of lit objects or our ears of vibrating objects. Below the threshold of sense perception of corpuscular aggregates there is a dimmer and cloudier impression of things as not yet sharply set out in distinct shapes and forms. Sense perception puts these latter into quantifiable focus but only by refining them out of a primary matrix of "qualitative" fuzziness. If our universe is a purposeful process, then it is not so much in secondary perception as in primary perception that it would give us an inkling of its teleological status. It is in the darkness of primary perception that our universe's aim toward value would first be vaguely perceived. Not science, but rather religious symbol and myth would have the task of bringing this sense of significance to expression. Scientific discourse, because of its distance from the universe as primarily perceived, cannot be the measuring stick for the trustworthiness of those symbols and myths pointing to a final purpose to events in nature and history. Science as it is usually understood is simply incapable of addressing the question of the possible purpose of nature since the material it deals with has already been abstracted out of the "qualitative" realm of value and placed in that of the merely quantifiable, subject only to mathematical calculation.

Scientifically oriented philosophy of nature has usually taken as unshakable the view of perception espoused by the empirical tradition whose charter members are Francis Bacon, John Stuart Mill, John Locke, and David Hume. The secondary and abstract nature of sense data has led it to view our primary perception of value, aim and significance in the universe as a mere "subjective" projection of our wishes onto the intrinsically neutral and valueless data of sense perception. Once we advert to the polar quality of perception, however, we need no longer view the sense of purpose as a projection, but rather as the cognitive feeling of something intrinsic to the universe entering into the depths of our primary sensitivity and seeking to come to expression in linguistic forms (symbol and myth) the validity of which cannot be assessed merely by scientific, quantitative criteria. But in order to accept this position we need first to reform our standard notions of perception; and this reformation will not occur without a conversion to the <u>radical</u> empiricism of attending once again to our experiencing itself. Further, the alteration of our inherited notions of perception cannot occur without a rethinking of the nature of physical reality and a radical critique of dualistic mythology.

Perception the Key to Causation

I am defending in this book the conviction that nature is something more than the meaningless, blind, absolutely unconscious process proposed by scientific materialism. But if nature is in any sense a purposefully oriented process, then it would have to be open to the influence of a transcendent, caring, intelligent principle. I understand the word "God" as pointing to such a principle. While I hesitate to identify the biblical Creator-God simply with efficient, formal or final cause of the universe, I do think that there is some validity in employing the category of causation analogously when we try to express the way in which God influences nature. I emphasize "analogously" because of the nonsensical implications resulting from a strictly literalistic transference to God of our mundane experience of causation when referring to the sense that our universe is ultimately cared for and impregnated with purpose. The term "influence" seems to me to be more flexible in its symbolic overtones; and the imagery of "flowing-in" suggested by this word fits the relation of God to world that we shall be developing in subsequent chapters. However, I shall continue to employ the term "causation" as well, even though I have some reservations about doing so.

An accurate understanding of <u>perception</u> is the key to our understanding the notion of causation.[6] The quality of physical reality that would render it open to being influenced is its pervasive perceptivity-mentality. The category of perceptivity that we apply to all occasions of actuality allows us to envision them as <u>actively</u> synthesizing the past into themselves. They are influenced in their own "subjective" feelings by the way they each uniquely receive and combine, suppress, or abstract elements of the world process that enter into the primary pole of each moment of perception. In this sense the fact of perceptivity makes causation a valid and intelligible notion.

In order to grasp more specifically how such causal influence might be possible let us examine more clearly the characteristics of the occasions that make up the world process. Each occasion owes its actuality to a "self-creative" process of feeling or "prehending" past occasions that may be called "objective" with respect to its "subjective" appropriation of them. The past, perished occasions are the "data" for synthesis by present occasions. Thus the past causally influences the present occasions by entering into them. But this causal influence is not to be understood according to the model of mechanical causation. It is not as though the present occasion is a totally passive recipient of the impact of the past, as one billiard ball is set in motion by another. Each present actual occasion actively and creatively synthesizes its past by conforming to it, but also by occasionally departing from the patterns of experience embodied by past occasions. The reason that the same rock, atom, or molecule can persist without change for millions of years is that its constituent occasions conform serially to each other without discernible modification as they synthesize their past. But there are also societies of occasions (live organisms for example) in which there is a pronounced element of non-conformity to the past. It is because of the possibility of synthesizing the past without necessarily completely conforming to it that novelty can make its entrance into the world-process. This capacity for partial non-conformity eventually allows life and consciousness to appear on a mineral landscape dominated by the conformity of entities to past patterns.[7]

Thus causality in a world of actual occasions can be understood in a non-deterministic way. The temptation to determinism in our thinking arises from the fact that the bulk of nature, the mineral level studied by geology, physics or inorganic chemistry is constituted by aggregates of occasions so conforming to their past that any present state in this inert realm seems to be the purely passive recipient of a series of events leading up to it. Present states or movements in the inorganic arena appear to us to be determined totally by the history of past commotion in the macroscopic order as described by classical physics. We are then inclined without warrant to apply this strong impression of mechanical causation to all of nature, even to the point of explaining life and mind as the passive, determined results of the aimless and blind movement of a dead and unconscious past.

The fact that, quantitatively speaking, most

of the societies of occasions making up the universe are dominantly conformal toward past patterns of experience does not entail that all societies of occasions are. The phenomena of life and mind, though quantitatively infinitesimal on a cosmic scale, attest to the potential for non-conformity that must be present even in the inorganic world in order that these novel patterns could have emerged in evolution. Entities manifesting vitality and consciousness are examples of societies of occasions that do not conform as rigidly to their past as do the occasions of a hydrogen atom or the Rocky Mountains. Thus our understanding of causal efficacy must take into account the apparent flexibility in nature, the apertures it leaves for the ingression of novelty.

In order to make sense of the fact of novelty in the world we must elaborate further the notion of "prehension". Prehension must be understood not only as the assimilation by each occasion of the past perished occasions. The notion of prehension must also be expanded to include the grasping of possibilities lying beyond those realized in the past. Without the occasion's prehension of further possibilities it would conform totally and completely to its past. If such complete conformity were the case, then determinism would indeed be the only feasible philosophy of nature. All causation would be, as scientism has usually assumed, blindly mechanical.

Even when prehension is dominantly conformal, however, it is not without an entertainment of new possibilities. Each occasion of experience is somehow open to a range of possibilities for synthesizing its past. In the act of prehension, however, we may say that it "de-cides," in the sense of cutting itself off from the many possibilities entertained, in order to realize only one set of such relevant possibilities. Prehension is a present creative act open on both ends, to the past on the one hand and to further possibilities or novelty on the other.[8] Causal activity in nature shares in this polarity. Each occasion is an _active_ "effect" synthesizing its causal elements creatively into its own subjective "enjoyment." It prehends its past into itself, thus allowing the past to influence it. But it does so only by simultaneously prehending a range of relevant possibilities in terms of which it de-cides to what extent it will conform to its past or advance beyond it. Causal influence enters into the self-constitution of each occasion, then, both from the past and from some principle of novelty, some source of possibilities, presenting itself for synthesis into each moment of experience.

Conclusion

Nature's openness to divine influence becomes intelligible only if we understand causation in terms of the perceptive or "prehensive" character of occasions. If there is a sustaining ground of order and a principle of novelty behind the evolving cosmos, then nature must be intrinsically open to such transcendent influence. Our task thus far has been to propose as consistent with modern physics, logic and the nature of perception that our universe does indeed possess this openness in a way that the abstract world of scientific materialism does not. We shall expand on our position in the following chapter by reflecting on the apparently _emergent_ quality of the cosmos.

Chapter IV

EMERGENCE

In traditional as well as in everyday habits of thinking we delineate at least four realms in nature: mineral, plant, animal, man.[1] Correspondingly we tend to ascribe distinct qualities to each: the mineral realm is thought of as inanimate, the vegetative as animate, the animal as sentient, and the human as conscious to the point of self-awareness. Evolutionary theory still maintains these distinctions, but it does so primarily in a temporal rather than a static sense. It hypothesizes that our universe has advanced in time from the inanimate, through the appearance of plant and animal life, culminating recently in the "emergence" of man with his capacity for language and reflective thought.

Our ordinary language would be thoroughly crippled if we did not continue to make these distinctions among hierarchically ordered levels or dimensions of cosmic phenomena.[2] Yet scientific theory often disregards the crisp demarcations our ordinary language and thought place at the boundaries of the inanimate, vegetative, sentient and conscious dimensions of nature. Science today sees no such clear lines anywhere. Beginning with the conviction that the inanimate world of subatomic particles and molecules described by physics and chemistry constitutes the basic construction material of the plant, the animal organism and the human brain, many scientific thinkers have questioned the "reality" of any other realm than that accessible to physics and chemistry. Physico-chemical analysis is unable to discern directly what people for centuries have referred to as life and mind. And so the latter are relegated to the status of "epiphenomena." As such, life and mind are given only a derivative being since they lack in themselves the hard reality of the objects encountered by physics and chemistry. Consequently, there is no need to draw the traditional lines of ontological discontinuity where there is utter material homogeneity, that is, where atoms and molecules cut across all the former boundaries in the hierarchy of beings.

We have already sketched the mythic, cosmological and epistemological background of contemporary attempts to explain life and mind completely in terms of insensate physical stuff. Dualism, scientific materialism and the empiricist doctrine of perception jointly constitute a compelling and nearly ineradicable tradition of thought to which many scientists unknowingly appeal. However, even apart from the tenuousness of this tradition as exposed by Whitehead's careful examination of it (only parts of which we have presented in the previous chapters), there are serious logical fallacies involved in its denial of the genuinely emergent character of life and mind. In the reduction of life and mind to atoms and molecules there lies an illogical maneuver of which Michael Polanyi has given the most devastating critique thus far.

Later in this chapter I shall provide a brief facsimile of Polanyi's critique. But for now, lest it appear to some readers that we are in dialogue with a phantom scientific ideal rather than with one that is seriously held, let us recall the famous statement of F.H.C. Crick, the celebrated Nobel-prize winning molecular biologist and author of the book, <u>Of Molecules and Men:</u>

> The ultimate aim of the modern movement in biology is in fact to explain <u>all</u> biology in terms of physics and chemistry.[3]

Crick's colleague, James Watson, goes even beyond this. He is convinced that not only heredity but other aspects of life as well are similarly reducible:

> Complete certainty now exists among essential-

ly all biochemists that the other characteristics of living organisms (for example, selective permeability across all membranes, muscle contraction, and the hearing and memory processes) will all be <u>completely understood</u> in terms of the coordinative interactions of large and small molecules.[4]

Occasionally scientific thinkers like physicist Gerald Feinberg go still further:

> If the physiological aspects of life are explicable in terms of physics and chemistry, it is likely that human mental processes are as well. Conceivably, the situation might be otherwise, and there might be some phenomena involved in the human mind that are not found elsewhere in the world. In that case it would be necessary to extend physics to include the new phenomena as well. However, the continuity of structure and function from nonliving matter to living and from the simplest forms of life to the most complicated strongly suggests that even the most characteristic human activities such as thought and consciousness have an explanation, as yet only partly known, in chemical and physical phenomena.[5]

I do not want to give the impression that the majority of scientists hold to this reductionist view. But since there are many influential ones who do, and since their opponents are at times unable to give reasons for their own opposition to reductionism, I think it is worthwhile to give our attention to the statements just quoted. It is important to do so ultimately because such assertions go hand in hand with the scientific repudiation of any teleological explanation. Accompanying the choice by many recent biologists to reduce their science to the level of physicochemical analysis there is a vigorous public refutation of any of their colleagues who persist in leaning toward teleology. Ernst Mayr of Harvard represents this stance:

> The proponents of teleological theories, for all their efforts, have been unable to find any mechanisms (except supernatural ones) that can account for their postulated finalism. The possibility that any such mechanism can exist has now been virtually ruled out by the findings of molecular biology.[6]

If teleology is imported into biology in the form of a "mechanism" then it deserves Mayr's chastisement. (His caricature of teleologists is a typical one). But the attempt by Mayr, Monod, Crick, Watson, etc., to explain life exclusively in molecular terms is no less reprehensible both for its naivete about the mythic, philosophical and epistemological tradition out of which it springs and for the deviations from logic in its "explanatory" procedures.

The Logic of Emergence[7]

It is not illogical scientifically to break down organisms and minds as far as possible into their physical coefficients. Such an analytic procedure is not only commendable but also necessary as a part of any adequate understanding of life and consciousness. Rather what is logically repugnant is the denial of any ontological autonomy, any distinct reality, to life and mind simply because these are not formally discernible objects of physico-chemical scrutiny. What is objectionable is the implicit metaphysics that bestows the status of "reality" only on atoms, subatomic particles and molecules but not on comprehensive wholes endowed with life and consciousness. In other words what finally cannot withstand the test of logical criticism is the rejection of genuinely emergent dimensions in nature.

Following Polanyi, I shall argue that emergent novelty and ontological discontinuity can enter into our evolving universe without in any way violating or disrupting the physical continuity that obtains at the molecular level. The denial of purpose in nature by scientism rests partly on the assumption that the discontinuities in world process are not real but only apparent. And if apparent emergence can be reduced to sheer physical resultance, then there is no need to ask questions about final causation or about any transcendent source of novelty.

Thus it is crucial that we focus our discussion here on the question of the ontological status, (that is the question of the "reality") of life and mind, the most obvious instances of allegedly "emergent" phenomena. Is it consistent with sound logic to maintain that these are mere epiphenomena fully explicable in terms of physics and chemistry?

It is very difficult for anyone familiar with modern science to dispute the evolutionary picture of our world-in-process beginning with relatively unorganized matter, moving gradually toward more complex atomic, molecular, cellular, vegetative and animal structures, culminating after millions of years in the evolution of man.

Our question, then, is whether physico-chemical analysis (in combination with some version of the theory of "natural selection") is by itself sufficient to account for this gradual evolution of plants, animals and man out of less complex organizations of matter. Or put in other terms, is the current methodological ideal of scientific atomism sufficiently broad to explain life and mind as we experience them? Is a totally blind, unconscious physicochemical process capable of producing vision and rational self-consciousness? Can an essentialy careless universe produce beings whose most admirable attribute is their propensity to care? Can a radically impersonal arranging and rearranging of molecules produce persons? Can a non-purposive movement of matter eventuate in beings whose very vitality depends upon their being animated by purpose?

Once dualism made it possible to siphon mentality out of the natural world, "matter" was left bereft of the perceptivity upon which alone cosmic purpose could be implanted. Then the stage was already set for the hypothesis that evolutionary emergence is at root impersonal and blind. The contemporary attempts to reduce biology to physics and chemistry are simply expressions of this very hypothesis.

The postulate that life originates purely by chance out of mindless and aimless shuffling of atoms and molecules is all part of our central question here concerning the logic of an emergent view of nature. Our entire discussion of this point may be focused on the question of the logic of the contention that biology and, by extension, neurophysiology are reducible to physics and chemistry. Crick's formula that biology is reducible to physics and chemistry is of central importance because, if it is logically coherent then emergence is indeed only an illusion, and the notions of final causation and purpose are dispensable in any intelligent attempt to understand nature.

Often today the biologist or neurophysiologist takes it for granted that all causation is mechanical. So when breakdown of a molecular process disturbs the functioning of an organism or of sentience and consciousness, he/she simply assumes also that molecular processes provide the full causal explanation for the <u>successful achievements</u> of organisms and minds. From the obvious fact that physico-chemical breakdowns cause the <u>failure</u> of performance by comprehensive wholes, the scientific atomist infers that physico-chemical processes cause their <u>success</u> as well. But such an inference is unwarranted, however attractive it initially appears. It employs the concept of causation in the same way in two entirely different situations, success and failure. Logically speaking, this is a category mistake.

It is of course undeniable that the breakdown of molecular processes in the cell, organism or brain does occasion dysfunction or even death. So the failure of the comprehensive whole does indeed follow upon the breakdown of prerequisite biochemical processes. Furthermore, successful functioning of a cell, organism or brain is contingent upon the recurrence of the most basic physico-chemical processes in their adherence to the laws of nature. Were there not a certain invariance about the way in which carbon atoms bond with others under identical conditions, or about the manner in which protein synthesis is charted and activated by nucleic acids, life would be impossible altogether. But is it logically correct to infer from these platitudes that physico-chemical processes explain the successful achievements of an organism simply because their breakdown leads to the organism's failure?

Let us illustrate the kind of logic employed here. I hope it does not seem presumptuous on my part to state that the writing of this book has involved what is usually called "mental" activity. In some sense or other it is the achievement of the mind of the author. (I shall leave it to the reader's charity to determine to what degree this may be so.) The use of reason, the appeal to experience and common sense, the shaping of theses and propositions, their formulation in sentences, paragraphs and chapters - all of these are the results of mental activity. If I have had any success in communicating meaning to the reader by writing this book, are physics and chemistry capable of completely explaining this achievement? The contentions of reductionist biology and neurophysiology lead us to expect that such explanation will eventually be forthcoming or can at least be provided in theory.

Of course, I would agree that in order to write a book successfully, the author's physiological and neurological apparatus must indeed be operating normally. The breakdown of this operation would occasion the failure to produce an intelligible work. Further, the physico-chemical elements and activities that form the substrate of the author's biological and mental acts must also be reliably patterned and programmed. Their avoidance of chaotic and whimsical jumbling is a necessary condition of such complex achievements as thinking and writing. But to call predictable biochemical processes a necessary condition of successful thought or writing does not make them exhaustive explanations. They are necessary but not sufficient conditions of the achievements of life and mind.

Physics and chemistry cannot completely explain the activity of writing a book, even in principle. No matter how thorough one's knowledge of the intricacies of physics and chemistry may be, this knowledge alone is incapable of providing the rules for successful writing. Physics and chemistry tell us nothing about how to utilize language, grammar and literary style. We do not go to the physicist or chemist (as such) to learn how to think out and write a book. We consult the literary experts, those devoted to the study of composition, syntax and writing technique. The physicist or chemist is simply not formally concerned with these issues.

The "heuristic field," the realm in which inquiry moves for the physicist or chemist is different from that of the literary critic. And the logical inability of science to formulate the rules for intelligent mental activity such as the composition of an essay, corresponds to the ontological discontinuity between the dimensions of reality analyzed respectively by the natural sciences and by literary criticism. The clue to the logical irreducibility of the latter to the former may be seen in the disparity of the questions we address to each level. By no stretch of the imagination or of logic does physics ask literary questions of style or chemistry questions of grammar. The material formally dealt with by literary criticism must therefore lie in an area inaccessible to exhaustive physico-chemical analysis. We may conclude then that mental activity is logically irreducible to the apparently insensate material analyzed by physics and chemistry. Questions of literary or intellectual success and achievement simply make no sense at the physico-chemical level.

Is Biology Reducible to Physics and Chemistry?

If it is questionable whether mental activity such as the planning and writing of a book can be fully explained by the sciences of physics and chemistry, it may not be so doubtful that life is also resistant to exhaustive explanation in terms of atomic and molecular analysis. But the same violation of simple logic is present in biological reductionism as we have seen in the case of attempts to reduce mental achievements. The sciences of physics and chemistry (or biophysics and biochemistry) can specify the atomic and molecular processes in the cell (as in its self-replicative "mechanism," the DNA molecule or in protein synthesis out of amino acids). But these sciences are incapable by themselves of defining what life is or even of recognizing it when it occurs. For such identification and recognition a logically distinct science, biology, is required. This science is based upon our human ability to recognize achievement in the biosphere.

Biology can be designated as a science logically distinct from physics and chemistry because its heuristic field is constituted by questions directed toward whole organisms (plants and animals), cells and their "achievements" rather than toward atoms and molecules as such. There would be no biology were there no such comprehensive wholes endowed with achievement oriented properties unspecifiable by the basic sciences. Scientific reductionism, however, wants to reduce biology to physics and chemistry, to explain the properties of "life", by thorough specification of the particulars (atoms and molecules) that are integrated into cells and organisms.

Let us again resort to an illustration in order to portray the absence of logic involved in this attempt at exhaustive reduction. Explaining life solely in terms of physics and chemistry would be analogous to explaining how a town got built simply by demonstrating the crafts of making and laying bricks.[8] These latter are of course a necessary condition for the successful construction of a town; and the conceptions of the architect and town planner are dependent for their implementation upon the competency of masons. But the character of the town cannot be apprehended even by the most meticulous examination of the processes of making and laying bricks. Operational principles are involved in building a town that cannot be grasped by a specification of the features of brick-work. We must also consult the architect and, even more, the town planner in order to understand more fully how the town came to be and what its true character and purpose are.

To hold that biology is reducible to physics and chemistry is no less absurd than trying to explain the building of a town exclusively by specifying the skills of making and placing building materials. Knowledge of biochemical processes is not coextensive with knowledge of life. Of course the successful achievements of organisms (adaptation and reproduction, for example) are impossible without the proper combinations of atoms and molecules in the genetic and developmental processes. But specification of these combinations says nothing about the possible organizing principles that "harness" biochemical processes and integrate them into hierarchically higher dimensions of being - life and mind.

Brick making and brick laying activities

leave themselves open to being controlled and ordered by the schemes of the architect and higher yet, the town planner. Without the purposiveness of the town planner and the architect, the brick laying process might still go on, but in an utterly random and haphazard way. Fortunately for the town planner, brick laying is a flexible enough process that it can be regimented so as to produce an endless variety of patterns corresponding to the planner's designs. Brick laying is not frozen into a routine so rigid that it mechanically and blindly runs its course impervious to any purposeful ordering by an extraneous principle of control. While the laying of bricks must fall within the limits of basic laws of physics and thus be determined "from below", it is indeterminate enough to be sequestered and styled by ordering agencies operating "from above."

However, if we look at a town merely from the point of view of the physical laws involved in masonry (laws concerning gravity, bonding of particles, evaporation, etc), we tend to focus on the determinism "from below" that is a necessary condition for brick laying. And while we are doing so we put in brackets consideration of the over-all indeterminacy in the brick laying process which leaves it open to being determined from above by a higher organizational principle. I am suggesting that perhaps the same sort of single-level approach is being taken today by those who reduce the science of life to molecular biology. Their gaze is so penetratingly fixed on the physico-chemical "building blocks" of life that they abandon any consideration of the openness of the whole chain of chemical reactions and atomic constituents to being harnessed by a hierarchically higher dimension whose focus is achievement and performance of skills. It seems entirely logical for us to hypothesize that the invariant physico-chemical reactions occurring in the cell (especially in DNA) leave themselves open to being ordered in a wide variety of ways by organizational principles the formulation of which is proper not to physics and chemistry but to biology.

The processes described by physics and chemistry are not so inflexible that they cannot be harnessed by higher, biotic principles in order to produce life and evolution. Without abrogating or even modifying the laws that determine bondings and pairings of atoms or molecules, principles pertaining to the biotic dimension can still control and orient physico-chemical processes in an extravagantly abundant variety of ways. In order to understand life we must consult those whose business it is to detect and formulate these higher organizational principles, i.e. the biologists. We should not expect, however, that these principles will be stated with the same precision and sharpness as the laws of physics and chemistry. For the principles of biology will have to deal with the elusive logic of achievement, whereas in the hard sciences there is no such experience as success or failure.

Continuity and Discontinuity

A common feature of most modern theories of evolution is their emphasis on the gradual, continuous movement of the universe toward organic complexity out of the more primitive phases of its unfathomable past. The absence of sudden qualitative leaps of matter from one kind to another (as depicted by geology, paleontology, comparative anatomy, embryology and other sciences) reinforces the gradualist hypothesis. And a physico-chemical perspective on evolutionary theory further blurs apparent discontinuities. The basic sciences are unable to discern any clear quantum leaps in the temporal-historical transitions from the non-living, through the living, to the conscious. So devotees of scientific atomism conclude to the ontological homogeneity of evolutionary phenomena. They look for simple mechanical causes that might account for the imperceptibly slow evolution of complex organic structures. And then they specify these "mechanisms" as the sole causal factors in the "emergence" of life and mind.

This line of inquiry and explanation may lead the micro-biologist to anticipate answers to the questions of life exclusively on one level, that of mechanical causation. Consequently, what starts out as biology may gradually drift over into a mechanistically understood physics and chemistry. The properly biological questions initially raised by our personal, empathetic encounter with the performances of whole living entities are edged out by preoccupation with the molecular constituents of life. The heuristic field is narrowed down; and a return is then seldom made to the question of what life is as we spontaneously recognize it in the achievements of animals and cells. Our intuition that life is ontologically distinct from non-living aggregates is strongly suspect as vitalistic and pre-scientific.

To those who profess this suspicion J.S. Haldane's reflections of a few decades ago on the uniqueness of biology would be embarrassingly out of date:

> That a meeting-place between biology and physical science may at some time be found there is no reason for doubting. But we may confidently predict that if that meeting-place be found, and one of the two sciences is swallowed up, that one will not be biology.[9]

It is not all clear today in the encounter of biology with the physical sciences that the digestive process runs in the direction Haldane predicted.

However, there is a sensible alternative to the reductionistic interpretation. Marjorie Grene, a disciple of Polanyi, clearly exemplifies this option. She maintains that it is possible for us to accept the causal physical continuity in nature's evolution while at the same time affirming an ontological discontinuity among the "levels" that emerge in the process:

> . . . to insist on epistemological and even ontological discontinuity is not to deny historical continuity, for conditions which are continuous can give rise to, or trigger, systems which once in existence are self-sustaining and hence not explicable entirely in terms of the conditions which produced them. . .
> The discontinuity of emergence is not a denial of continuity but its product under certain conditions.[10]

Denial of the reality of emergence pivots on an inability to hold together logically the concept of temporal-historical continuity with that of ontological discontinuity. And yet there is no logical incoherence in thinking of nature as a hierarchy of distinct dimensions integrating a continuous, unbroken chain of physico-chemical occurrences, (just as the architect's designs do not interrupt the continuity of the brick laying process, but simply impose a determinate structure onto it.) In fact allowances can also thus be made for the role of chance in the emergence of life and in the mutations that are required for the evolution of new species.

Our hypothesis is that the evolving universe is a field of ontologically distinct patterns of ordering principles that may be released and become "incarnate" as the result of random triggering circumstances but that cannot be adequately accounted for simply by the specifying of such circumstances.

A simple picture may illustrate the compatibility of a historically continuous set of triggering circumstances (involving randomness) with the emergence of entities governed by such novel ordering principles. Suppose a flame is accidentally triggered by the friction of two highly combustible materials.[11] In a simplistic sense we may say that the chance rubbing together of the materials explains why the flame appears. Or, in a scientific context, we may insist that the intense release of kinetic energy produces sufficient heat energy to ignite the combination of chemicals. In either case, however, we are still talking merely about triggering causal events. We are not yet defining what a flame is. In order to define more fully what a flame is we would have to go beyond the mechanical or physicochemical description of its origin. We would have to recognize that it is an open system sustained by a continuous input of gases from the environing atmosphere and from the burning material, and giving off heat energy into the environment. Without a surrounding field sustaining it there could be no flame at all. Once the flame appears on the occasion of random triggering circumstances its nature and activity are not rendered fully explicable simply by specifying the chain of occurrences leading up to it. The "field" in which it appears must also be taken into account. Within its environing field the flame possesses a stability, a constant though to some extent variable form, open to the influx of atoms and the steady outflow of energy and waste material. This constancy implies that there are ordering principles in its field stabilizing the flame that were not present in the triggering circumstances that produced it. To understand a flame then we must also posit the existence of and try to formulate these "ordering principles" as well.

Polanyi insists that "it is a fundamental property of open systems . . . that they stabilize any improbable event which serves to elicit them."[12] And since live organisms are open systems "the first beginning of life must have likewise stabilized the highly improbable fluctuation of inanimate matter which initiated life."[13] Randomness is somehow domesticated by these open systems. Therefore we must postulate the presence in our emergent universe of a field of organizing factors, extraneous to triggering circumstances, that allow new types of order and functioning to burst forth and to persist.

In a sense roughly comparable to our picture of a flame it is possible (without in any way abandoning the notions of purpose and emergence), to hold that life also appeared by "chance." It is entirely possible that a random triggering event or series of events (lightning or some form of electricity charging a primordial soup of methane, ammonia, etc.; bombardment by cosmic radiation and other hypothetically random occurrences) was the occasion through which a whole new dimension, that of the biosphere, flooded onto the terrestrial scene. Understanding what life is, however, requires more than hypotheses about the triggering circumstances that released it in the first place. In order to know life we must become aware (tacitly at least) of the surrounding field of life-stabilizing factors and of the the phenomena of achievement toward which all life and evolution are oriented. The fact that forms of life can succeed or fail (in the performance of a skill, in the discovery of a suitable environment, in reproducing, in warding off death, injury, etc.) places them at a different level of being from purely physical processes to which the terms success and failure are inapplicable. (How can any chemical reaction be called a success or failure in itself?) Consequently, knowing life means becoming aware of the ordering principles that direct it towards its achievements and that give it stability.[14]

We have been maintaining that physics and chemistry logically cannot deal with these ordering principles. Are we therefore to deny the reality of such principles? Or are we not rather called upon simply to relativize the cognitive prowess of physical science? If we take the latter approach our universe would no longer necessarily appear alien to the operative presence of purpose. How such purpose might specifically manifest itself in out emergent universe, though, is the subject of the following chapter.

Conclusion

I cannot end this chapter without emphasizing that there are respected biologists who take exception to scientific reductionism. Sir Alistair Hardy, I am sure, speaks for many:

> I shall shock some of my colleagues when I say that I feel a sympathy for Shaw's elderly gentleman in Back to Methuselah who said, "they tell me there are leucocytes in my blood and sodium and carbon in my flesh. I thank them for the information and tell them there are black beetles in my kitchen, washing soda in my laundry and coal in my cellar. I do not deny their existence but I keep them in their proper place." We must keep physics and chemistry to their proper proportions in the scheme of life.[15]

And Barry Commoner issues a forceful warning about the eventual environmental consequences of the methodological program of resolving biology into physics and chemistry:

> I sometimes think that the difficulties we now face in controlling water, air, and soil pollution, and the undue dissemination of radioactive materials, are the result of a common impression that "the boundary between life and non-life has all but disappeared." In fact if we do not mend our ways the statement may, after all, turn out to be true.[16]

What Commoner is suggesting is that the reductionist agenda has more than purely theoretical implications. If we really start believing that life is reducible to the inanimate, it might not be long before we have foisted this "ontology of death" onto the actual world itself. In this chapter, at any rate, I have argued that reductionist methods contradict the most basic elements of simple logic.

Chapter V

PURPOSE

Once we view nature as a hierarchy of emergent dimensions in which physical reality is pervasively "experiential," what can we make of the question of purpose? Is emergence a trend in nature leading toward some goal that we can clearly anticipate? Is it an essentially aimless play of cosmic forces signifying nothing? Or is it a process whose possible direction can never be fully clarified, but which may nonetheless be called purposeful, significant and meaningful?

If there is purpose in our emergent universe, we may now legitimately speculate that it would be present in the form of a "higher" or "deeper" dimension influencing and ordering the lower (or surface) dimensions. We have already established that the biotic principles elucidated by the life sciences can order physico-chemical processes without violating them. And it is evident that mental operations, following principles logically irreducible to bio-chemistry, can impose an even higher order on physical and biological processes without disrupting the causal continuity of the latter. An unbrokenness at a lower "level" does not rule out, but rather makes possible its integration into a higher one. Is it possible, then, that there is yet a further, transcendent integrating and ordering influence operative in nature, one that "orders" the lower dimensions including the noosphere without disrupting their apparent continuity? Such an hypothesis, though not scientifically verifiable, is at least consonant with the logic of emergence.

Experience and logic both yield the principle that a higher dimension can comprehend a lower, but not vice-versa. Biotic principles can integrate, order or "comprehend" molecular occurrences, but the latter cannot do the same to the former. Mental activity can embrace and rely upon the particulars of biological processes and biochemical reactions in the nervous system, but our grasp of the workings of the lower processes does not by itself give us an understanding of the nature of consciousness. Comprehension is unidirectional. Consequently, if the lower cannot comprehend the higher, and if there is an ultimate, transcendent dimension encompassing and influencing nature, then it could not be comprehended by our own minds in any case. In other words, if there is a divine scheme of purposefulness that envelops and grounds the dimensions of our emergent cosmos, we would not be able to grasp it in an objectifying, controlling way. Rather it would grasp us, and we would then experience ourselves (initially in what we have called primary perception) as being taken up into such an ultimate synthesis.

There have always been available in human life mysterious modes of expression intimating such a sense of being embraced by a deeper dimension. These expressions are especially, though not exclusively, the symbols, myths and rituals of religion. Unable to fit these elusive expressions into an objectifying, scientific understanding of the world, we often dismiss them as illusory. Yet this derogation of the language of religion as incongruous with what we know scientifically about a one-dimensional universe, may turn out to be inappropriate in terms of an emergent, multidimensional one.

Faith

In an emergent universe it is never possible, standing within a lower dimension, to reach an objectively adequate knowledge of the organizational patterns that may be operative at a "higher" one. All we can say with certainty is that the reliable functioning of the lower is a necessary but not sufficient condition for the sucessful performances of the higher, and that the breakdown of the lower may bring about the failure of the higher.[1] But knowledge of the lower, no matter how sophisticated, cannot account for the higher's successful achievements. To account for these achievements we need to have a knowledge of the extraneous ordering principles that harness lower processes and integrate them into novel arrangements. The science of the lower cannot give us such principles. Thus we need a hierarchy of "sciences" or modes of knowing corresponding to the hierarchy of emergent dimensions in nature. The science of the lower is not "adequate" to comprehend the organizing principles that pertain to a higher emergent set of occurrences. Whenever, as is often the case today, physical science attempts to encompass the totality of the universe, the higher (or deeper) is inevitably diminished, confined to the heuristic field of the lower (or surface) mode of inquiry. And within such imprisonment the world's genuinely emergent dimensions are lost, or their existence even denied.[2]

However, there is a mode of consciousness by which we at the human level of emergence might become aware, in a non-controlling, non-comprehensive sort of way, of possible higher purposive principles and patterns of influence operative in the cosmos. Certainly any controlling type of cognition is ruled out in principle. But what is not necessarily excluded is another kind of cognition by which we <u>leave ourselves open</u> to being grasped by whatever <u>higher or deeper</u> ordering influence there may be. The classical, but often misunderstood, name for this cognitive stance of <u>receptivity</u> to new possibilities of being comprehended is <u>faith</u>.

Faith is a problematic term in our science-dominated intellectual world. It often seems to mean the attitude of those who are afraid to look at the facts or who have no mature interest in the real world. However, I would like to argue that faith may be understood as a reality-probing stance of consciousness if taken in the context of emergence. Faith is simply a confident receptivity to and active appropriation of new possibilities of emergent order. As such it has its roots in the cosmic process itself. Faith is the route evolution takes at the human dimension of emergence as the universe ventures into the future. In order to clarify this notion of faith, however, we have to set it apart from those impressions many intelligent people have that it is a groundless commitment to absurdities or a clinging obsession with certitudes that may never be challenged. Instead, in a radical sense of the term, faith means an adventurous and exploratory rather than a strictly dogmatic posture. And its orientation toward reality consists precisely in its trustful openness to the reception of novel forms of order. In this sense, then, it is through faith that we would become aware of nature's purpose, should there be such.

Let us explore further the role of faith in the context of an emergent universe. As Polanyi has brilliantly demonstrated, each lower dimension in a hierarchy of emergents, without violation of its own internal structure, leaves itself open to being ordered by a higher set of principles extraneous to itself. Without this stance of openness the lower would be impervious to any integration into the higher, and so there would be no possibility of creative advance in the universe. But the most characteristic way in which the lower process leaves itself open to the higher is to function without violation of its own internal principles and structures. No suspension of the laws operating at the lower level is required for a novel emergent possibility to make its appearance in evolution. No exceptional performance is needed for the emergence of a deeper and more complex dimension. In fact there must be a consistency and predictability at the lower level in order for the emergent pattern, the new integrating reality, to appear and to function.

If for example there were a breakdown at

the level of the DNA molecule's physico-chemical constituents and reactions, the cell's transmission of essential genetic information would be frustrated, and life would not be given an opportunity to appear, let alone perform and achieve. For life "dwells in" and "relies upon" the consistency of the physico-chemical processes it integrates into its biotic configurations.[3] Or, to give another example, if neurological routines were disturbed, then the performance of mental operations would likewise be impeded. So in order for them to fulfill their role in emergence, the lower or subordinate processes need only function normally, predictably and continuously. We shall be able to draw some important consequences from this postulate later on.

Evolution has now advanced into the "noosphere," the domain of man and consciousness. This latest dominant emergent dimension almost daily seems to be taking on an increasingly planetary aspect. Teilhard de Chardin has expressed this impression with an incomparable vividness:

> All round us, tangibly and materially, the thinking envelope of the earth-- the Noosphere--is adding to its internal fibres and tightening its network; and at the same time its internal temperature is rising, and with this its psychic potential. These two associated portents allow of no misunderstanding. What is really going on, under cover and in the form of human collectivization, is the super-organization of Matter upon itself, which as it continues to advance produces its habitual, specific effect, the further liberation of consciousness.[4]
>
> . . .
>
> Whether we like it or not, from the beginning of our history and through all the interconnected forces of Matter and

Spirit, the process of our collectivization has ceaselessly continued, slowly or in jerks, gaining ground each day. That is the fact of the matter. It is as impossible for Mankind not to unite upon itself as it is for the human intelligence not to go on indefinitely deepening its thought! . . . Instead of seeking, against all the evidence, to deny or disparage the reality of this grand phenomenon, we do better to accept it frankly. Let us look it in the face and see whether, using it as an unassailable foundation, we cannot erect upon it a hopeful edifice of joy and liberation.[5]

It is not my purpose in this book to argue for the legitimacy of Teilhard's particular version of the universe's purpose, even though I am attracted to it in many ways. Instead I envision this volume as a prolegomenon to the sort of speculation that Teilhard has undertaken. I see a preliminary necessity for demonstrating the theoretical congeniality of nature to any kind of teleological interpretation in the face of the many contemporary denials of such a possibility. Teilhard himself has not provided an adequate theoretical discussion of the notions of physical reality, perception, causation and emergence upon which to implant his evolutionary teleological vision. And for this reason I have made appeal to Whitehead and Polanyi who are more comfortable with the complexities of philosophical discussion than is Teilhard.

Nonetheless, I see in Teilhard an openness to further emergent possibilities that exemplifies what I am calling faith. He opens us up to the possibility that the noosphere, dominated by the phenomenon of intercommunication among personal, cognitive centers may constitute a new "lower" dimension that could conceivably be taken up into a "higher" one. His thinking seems to be consistent with the logic of emergence; and if there is present in the universe any deeper or higher ordering principle, it is through the kind of visionary consciousness Teilhard exemplifies that we would be given a sense of it.

As Teilhard insists, there is no reason for us to think that the present status of the noosphere constitutes the end of the evolutionary process. In fact, when viewed by cosmic standards, beings endowed with the faculty to communicate linguistically, to express their ideas and aspirations, and to form communities around shared hopes have not been inhabitants of the terrestrial portion of the universe for more than a flash of time. The human race, in other words, is possibly very early in its development and is by no means clearly the climax of cosmic emergence. If we are to speculate, with Teilhard, about the future of the evolutionary trajectory, therefore, we may at least conjecture that it will not deviate from the logic of emergence that dominates episodes preceding us. Thus we would express our continuity with the evolutionary past by leaving ourselves and our humanity open to being synthesized into increasingly deeper organizing influences. When I use the term "faith" I am referring to this stance of adventurous, risk-filled openness that would allow such a possibility to take hold of the noosphere without in any way disrupting its "normal" forms of interaction.

In a qualified sense we might be able to see in mankind's religious longings the primordial, but by no means the exclusive expression of faith. Whether explicitly religious or not, however, the only cognitive posture that would be "adequate" to the presence of a teleological dimension in nature is faith, and not science exclusively. In searching for any evidence of purposive influence, the scientific approach tends to look for instances of discontinuity in natural processes whereby a teleological push or pull would insert itself somewhat obtrusively into the fabric of nature. (See the statement by Ernst

Mayr quoted earlier for an example of the scientist's expectation of an exceptional display of purposive presence as a condition for his accepting a teleological perspective.)[6] But not being able to verify such instances of disturbance of physical continuity, scientific reductionism rejects the possibility of any teleological presence whatsoever. If there is a higher purpose in nature, scientific thinkers often expect that it would somehow intrude into the lower spheres accessible to scientific inquiry. They do not grasp the fact that such an intrusion would be a violation of the laws and logic emergence: the higher dimensions never interfere with or suspend the workings of the lower. We have insisted all along that unbrokenness at the lower level is a condition of and not an argument against the presence and effectiveness of higher organizing factors. Consistent with this principle we are now maintaining that no magical, extraordinary interruption of physico-chemical, biological, psychological or interpersonal transactions is required for us to accept the reality of a transcendent ground of emergent order. The manifestation of such a presence need not take place either in violation of or apart from nature or human nature in particular. Rather it could occur in complete continuity with the logic of emergence. It need not interrupt the normal flow of human interpersonal encounter any more than the building of a town interrupts the laws operative in the juxtaposition of brick and mortar.

We may conclude, therefore, that the notion of purpose is logically compatible with the emergent aspects of nature but that the discernment of such purpose cannot be achieved by a purely scientific quest for peculiar mechanisms or unusual disturbances of the normal, lawful flow of physical, biotic and conscious activities. Rather, the sense of a deeper, purposively ordering presence in nature would reside in the non-scientific mode of consciousness that we have called faith.

Purpose and Physical Reality

We have attempted thus far in this chapter to argue for the theoretical concordance of transcendent purpose with the logic of emergence. But we may also understand "purpose" in terms of the "mental-experiential" character of physical reality that we set forth in the earlier chapters. Throughout this book we have been firmly maintaining that the reconciliation of the notions of nature and purpose can occur only after the view that physical reality is non-mental has been challenged. A mindless universe is hardly conformable to any truly teleological interpretation. Certainly ancient and medieval thought were aware of this axiom. The Greeks, for example, saw nature as saturated with mind and, therefore, with meaning. And traditional philosophies have often regarded our own individual cognitional faculties as microcosmic instances of a macrocosmic mind or logos that runs throughout the purposive universe. Today, however, we tend to condense mentality into our own individual cerebral frameworks, and to imagine everything else as devoid of any semblance of mentality.

We have criticized the conventional materialistic picture of physical reality, perception, causation and evolution that follows from a thorough-going expulsion of mentality and feeling from the fundamental constituents of nature. It remains now for us to explore the idea of purpose in relation to our alternative notion of physical reality. What precisely do we mean when we speculate that this experiential universe is purposive?

To begin with, we do not mean that there is a predetermined direction to world process. Both scientists and theologians have proposed goal-directed (orthogenetic) interpretations of evolution in the past. And these interpretations have been justifiably criticized as simplistic and naive by modern biologists and physicists. The presence of entropic trends, of chance and indeterminacy in physical reality is too obvious for us to hold that nature is the deterministic implementation of some blue-print rigidly inscribed in the nature of things by God or any other imaginable cosmic principle. Strait-jacket teleology has been legitimately scorned. There simply is no scientific or religious evidence of any pre-established cosmic plan.

As an alternative to such a disreputable point of view, however, we might understand purpose in the universe in terms of what has sometimes been called a "loose" teleology.[7] By this I mean that the cosmos may be a significant, <u>value-laden</u> process without needing to be strictly directional in its advance through time. The events that make up world process may be rescued from seeming oblivion and insignificance without corresponding to any deterministic conception of a goal for cosmic becoming. "Telos," in other words, need not entail a specific "finis."[8]

The notion of purpose comes to us in part from our experience of historical existence. Human concern for meaning has for centuries sought intelligibility primarily in the context of social and political events that constitute history. This longing for meaning has given rise to preoccupation with purpose in socio-historical life. And out of this concern there has developed the question we have been discussing, whether nature itself, in its sub-human as well as in its human dimensions manifests any caring, providential influence. The concerns expressed throughout this book are not derived from any nakedly unhistorical encounter with nature, since this is not a possibility in any case. Rather they erupt out of the questions and uncertainties of an entire historical period. The question of nature and purpose would not arise in the first place, nor would it have any interest to us, apart from our own unique historical situation with its own particular hopes and threats.

The notion of purpose derives also in part from our experience of machines and other artifacts invented by man in order to realize specific functions. "Purpose" is a hybrid term with various levels of meaning, one of which is at times a quasi-mechanistic one. In a machine dominated age we tend to scrutinize the natural world for patterns that we are familiar with from observing the products of engineering and cybernetics. We look for mechanical explanations, feed-back processes and other trends that might make aspects of nature intelligible to us in terms of our own functionally purposive intelligence. And we are often successful in verifying the presence of such teleonomic structures in discrete phenomena. But whether we can find and verify any analogous functional scheme in the universe as a whole is another question altogether. Indeed it seems that if the universe is purposeful, it would have to be so in some other sense than we find in the mechanical and historical fields of interest.

When we ask whether nature as a whole exhibits any purpose we notice immediately that the question is thwarted by the lack of perspective we have on any possible universal patterns and by the vast tracts of time and space over which nature sprawls in its depth of unavailability to our direct experience. We simply cannot give an answer anything like that we seek on the smaller scale of teleonomic performance of machines, computers and organisms. Regardless of how we understand purpose, then, if we hold that nature is influenced by any sort of final causation, this contention will appear to many to be no more than a stab in the dark without any empirical warrant.

"Purpose," though, need not be restricted to meaning goal-directed activity, whether historical or mechanical. There is a deeper sense in which we can understand the term so that in spite of our lack of universal perspective we may still attribute a teleological aspect to the cosmos. We may understand the purpose of world process in terms of the notion of <u>value</u> interpreted in an aesthetic sense.

Purpose as Aesthetic Value[9]

The term "purpose" cannot be grasped apart from the notion of value. Only orientation toward value renders a movement purposeful. So purpose will be understood here simply as the defining quality of any process aiming toward the realization of value. But if we are to get at its roots, value needs to be understood aesthetically. Aesthetically interpreted, value entails a synthesis of richness with harmony, complexity with order, novelty with continuity, and intensity with stability. Above all, aesthetic value implies the transformation of contradictions into contrasts that arouse a fullness and intensity of feeling, a sense of beauty, in those who experience the aesthetic object. We spontaneously value those entities that combine the polar qualities just mentioned more than we do those things that are relatively simple in their make-up. The human brain, for example, is granted a value that is lacking in a lump of clay. If we are to formulate a reason for the disparity of the respective evaluations, we can do no better than point to the degree to which each integrates complexity into an overall harmony. The human brain combines complexity with organization more intensely than does a lump of clay. A great work of art, to give another example, is able to integrate a multiplicity of fine contrasting shades of sound, color or verbal meaning into an overall composition of balance and proportion bearing an intensity that is lacking in a work of lesser stature. It is the combination of nuance with harmony that evokes aesthetic appreciation. Nuance without harmony and harmony without nuance both fail to arouse our aesthetic sensitivity and valuation. Value, therefore, is a quality inhering in the patterned unification of elements that in terms of an alternative frame of reference may be irreconcilable or contradictory. In a pattern of value, conflicts are turned into satisfying contrasts by the overall harmony of the aesthetic frame of reference.

It is in this aesthetic sense of value that we may understand the notion of cosmic purpose. Our universe can be understood as an aesthetic reality unifying contradictions into a harmony of contrasts that we might pronounce beautiful and, therefore, good, significant, meaningful.

We may grasp this aesthetic notion of value more firmly if we contrast it with the notion of evil, the contrary of value.[10] Evil is a quality associated with trends, persons or phenomena that remain in or degenerate into chaos or triviality when the possibility of harmony and intensity is in fact open to them. For example, war is evil as long as the possibility of peace is open to the combatants. Environmental disintegration is evil when programs could be implemented to enrich and balance complex natural processes. That is, evil may be understood in terms of destruction, disorder or chaos. But "evil" may signify <u>unnecessary triviality</u> as well. If I am capable of enriching my experience, expanding my knowledge or intensifying my sense of beauty, and yet refuse to do so even though opportunity is available, then I am siding with triviality. In a processive world-view the option for monotonous triviality would be an option for evil inasmuch as it turns aside from the pursuit of value, a pursuit that aims for the continual enrichment and intensification of physical reality, consciousness and life. On the other hand if I precipitously and ineptly venture to grasp too much at one time I might be overwhelmed by the complexity of the material that I attempt to appropriate. In this case I risk intellectual, emotional or spiritual chaos. I might try to bite off too much and in the process go mad, or at least become confused. Such disorientation, lack of harmony, is also an evil, though its degree varies considerably from case to case. Discord and unnecessary triviality are the central components of evil situations.

Purpose, therefore, would be the quality of any physical, mental, social, historical or natural process that aims beyond triviality and chaos toward maximizing harmony and intensity. No predetermined goal is required for the evolving, emergent cosmos if we understand its purpose in this aesthetic sense. Its aim toward beauty is the teleology of cosmic process.[11]

In conceiving of nature's purpose in an aesthetic sense, we have to recall again and again that our human experience lies on a continuum with the rest of nature. Even though the intensity, scope and originality of our experience differ in degree from those of non-human occasions, nonetheless, there is a structural similarity in all modes of cosmic feeling or prehension. Each occasion feels its past, anticipates novel possibilities and fuses these together in a momentary "enjoyment" or affective intensity that we can legitimately refer to as aesthetic in nature. What occurs in each moment of cosmic process, therefore, is not totally different from the events of sensitivity in artistic creation. Thus our humanly conscious creation and appreciation of beauty exemplifies the kind of aesthetically unifying experience that is pervasive throughout the universe.

In the creation of a work of art, such as a painting, for example, the artist, selecting from innumerable possibilities, fuses into a novel unity many diverse patterns, contrasting shades of color and lighting, subtleties of positioning, variations of theme, texture and emphasis. The aesthetic value of the painting will be proportionate to the degree of intensity to which variety, diversity and contrast are gathered together into novel unity. If contrast is too obtrusive, then the over-all harmony of the painting is jeopardized. In that case disharmony would inhibit aesthetic enjoyment. But if there is inadequate contrast and variety, then harmony would be so pervasive as to be monotonous. Aesthetic intensity would be negligible, and the painting would be trivial and unenjoyable.

The "purpose" of the artistic project of painting a picture is to realize the highest possible relevant integration of variety with harmony. The more intense the harmony, the richer is the aesthetic enjoyment. The artistic intention of combining variety and harmony may require that a portion of the painting, taken in isolation from the whole, will exhibit a fragmentary note of discord. But a perspective that takes the whole painting into account may be able to unify these discordant particulars into a wider and more intense harmony of contrasts. In this broader perspective, therefore, what is disharmonious from a limited point of view actually contributes to the aesthetic enhancement of the whole. In an aesthetic pattern contradictions and clashes are transformed into contrasts that heighten the value of the totality. May we not envision cosmic evolution as a process of aesthetic unification of often temporarily irreconcilable aspects into a "creation" of unimaginable beauty?

In reading a novel, to use another example of artistic creation, we often feel discomfort while immersed in the conflicts and crises of characters introduced by the novelist. Reading about specific problems besetting them momentarily reproduces in our own feelings the elements of discord presented in the narrative. If we were to cease reading the novel at these critical moments, the conflict and discomfort induced in us would remain suspended and unresolved. But if we continue reading and feeling these episodes in the light of the totality of a well-wrought novel, then the temporary uneasiness provoked by particular fragments of the work may contribute in their uniqueness to the overall aesthetic enjoyment of the literary creation. In retrospect our enjoyment of the whole novel depends upon our having dwelled within the particular episodes which, felt in isolation from the rest of the narrative, are often absurd, unintelligible. The novelist's purpose is to avoid both triviality and discord by weaving a relevant variety of detail into an over-all harmony that gives pattern and significance to otherwise mutually discordant incidentals. Of course no novel is ever perfect, and so it tends to some degree toward either triviality or disharmony. The perfect balance is never achieved in actuality. Perfection, understood as the ideal unity of harmony and intensity is the aim of cosmic process, not the achievement of individual aspects or phases within the process.

These examples of aesthetic creativity taken from the realm of human experience should not be disengaged from their cosmic setting. Both in its individual occasions as well as in its totality our universe may be viewed as such an aesthetic process. The fusion of novel possibilities into various modes of harmonized intensity that we experience in human creativity is representative of what goes on throughout cosmic process. The dynamism of the universe may, therefore, be understood as an aim toward the highest possible attainment and enjoyment of beauty. Such an aim would suffice to imbue our universe with the purpose we seek.

Conclusion

From the very limited vantage point that each of us occupies within the emerging universe, discord often seems to be dominant over harmony. We are often even inclined to take our individual experiences of tragedy as the key to the whole universe. However, the aesthetic model of cosmic purpose suggests that our own experiences may be lacking in perspective. There is perhaps a perspective on the universe that we do not ourselves have, but which would be able to unify into an aesthetic whole even those contradictions and absurdities that we deem most insurmountable. I think the word "God" may in part be understood as pointing to such a perspective. In the following two chapters I shall develop this idea in reference especially to the facts of perishing and evil.

Chapter VI

PERISHING

According to the ideas of emergence and value that we have presented there is a fragility in the universe that appears in direct proportion to the degree of intensity and complexity of emergent phenomena. That is, things of value are afflicted with an instability or precariousness that might make us question their value after all. Entities that derive their aesthetic intensity from the manifold nuances and contrasts that they integrate and rely upon are not guaranteed an indefinite prolongation through time. There is always the threat that the complexity of such emergent value-laden beings will overwhelm their harmony. Complexity may persist indefinitely, as may trivial modes of unity. But harmonized, organized complexity is itself eminently perishable. Patterns of physico-chemical activity, for example, may never deviate from strict routines, but the harnessing of these invariable patterns by higher dimensions into animate or conscious organic structures is precarious. Such "intense" structures are incapable of enduring beyond a brief span of time. The emergent phenomena, life and mind, are obvious illustrations of this fleetingness. There is everywhere the threat of instability to any intense integration of complexity and harmony. Along with a general aesthetic enrichment of the universe by the emergence of life and consciousness there has appeared a fragility that is constantly exposed to the entropic tendencies of physical reality. The precariousness of things seems to increase in direct proportion to their preciousness.

We must learn to think of cosmic purpose, if we are to think of it at all, in terms consistent with the evanescence of occasions and cherished societies of occasions. If our notion of purpose is to be faithful to the facts we cannot disregard the most obvious one of all: things perish.[1]

Anything of value must, of course, have some quality of endurance or else we would not be able even to recognize it, let alone revere it. But the things we treasure the most such as life, consciousness, personality, moral goodness, heroism, culture and peace all abide only tenuously. They are seemingly tinged with a kind of "unreality" that makes them appear epiphenomenal. While they are the most important things to us, their constantly passing away makes us wonder just how real and significant they can possibly be in the final analysis.

Our anxiety about the meaning of our own lives as well as that of the universe as a whole generally arises simultaneously with our experience of the transience of the most deeply valued entities. Things perish, including those we hold most dear, and so they apparently fail to make a mark of enduring significance. The fact that cells degenerate, that organisms decay, that our own lives ebb toward death, that civilizations eventually fall and that noble deeds and ideals fade into oblivion--all this makes us wonder how the universe could conceivably have any abiding seal of purpose. Can there be purpose without some aspect of permanence to the flux?

Without some sense of the everlastingness of the value achieved in the emergence of nature we might easily concur with the dour ruminations of those ancient and modern writers who have voiced an anguished pessimism as a result of their sensitivity to impermanence. Marcus Aurelius, for example:

> Time is like a river made up of the events which happen, and a violent stream; for as soon as a thing has been seen, it is carried away, and another comes in its place, and this will be carried away too. (<u>Meditations</u> IV, 43)

Do not consider life a thing of any

> value. For look to the immensity of time behind thee, and to the time which is before thee, another boundless space. In this infinity then what is the difference between him who lives three days and him who lives three generations? (IV, 50)
>
> All things are changing: and thou thyself art in continuous mutation and in a manner in continuous destruction, and the whole universe too. (IX, 19).[2]

Or the tragic prospectus laid out by the author of Ecclesiastes:

> The hearts of men are full of evil; madness fills their hearts all through their lives, and after that they go down to join the dead. But for a man who is counted among the living there is still hope: remember, a live dog is better than a dead lion. True, the living know that they will die; but the dead know nothing. There are no more rewards for them; they are utterly forgotten. For them love, hate, ambition, all are now over. Never again will they have any part in what is done here under the sun. (9, 3-6 New English Bible)[3]

The reigning philosophies of nature, influenced as they still are by the scientific materialism of the classical era in physics, are incapable of sustaining any hope that things of value somehow escape being utterly forgotten. William James has shown with deep feeling how impossible it is to reconcile materialism with any human longing to rescue permanence from the stream of passing events:

> That is the sting of it, that in the vast driftings of the cosmic weather, though many a jewelled shore appears, and many an enchanted cloud-bank floats

away, long lingering ere it be dissolved—even as our world now lingers for our joy—yet when these transient products are gone, nothing, absolutely <u>nothing remains</u>, to represent those particular qualities, those elements of preciousness which they may have enshrined. Dead and gone are they, gone utterly from the very sphere and room of being. Without an echo; without a memory; without an influence on aught that may come after, to make it care for similar ideals. This utter final wreck and tragedy is of the essence of scientific materialism as at present understood. The lower and not the higher forces are the eternal forces, or the last surviving forces within the only cycle of evolution which we can definitely see.[4]

The victory of the lower over the higher forces seems to be most decisive in the case of human mortality. It is the dreaded vanishing of our own personalities in death that stirs us to the greatest anxiety. Paul Tillich thinks that ". . . in the depth of the anxiety of having to die is the anxiety of being eternally forgotten."[5] Man was never able to bear the thought of having his being thrust into a past where it would be totally lost to memory. And so humans have always sought ways of resisting the fact of their perishability.

. . .the Greeks spoke of glory as the conquest of being forgotten. Today, the same thing is called "historical significance." If one can, one builds memorial foundations. It is consoling to think that we might be remembered for a certain time beyond death not only by those who loved us or hated us or admired us, but also by those who never knew us except now by name. Some names are remembered for centuries. Hope is expressed in the poet's proud assertion that "the traces of his earthly days cannot vanish in eons." But those traces, which unquestionably exist in the physical world, are not we ourselves, and they don't bear our name. They do not keep us from being forgotten.[6]

Thus for ages people have asked: "Is there anything that can keep us from being forgotten?"[7] Is there anything that might guarantee that nothing real is every totally pushed into the past? Affirmation of purpose has always required some positive answer to these questions. Unless perishing is less than absolute, unless transience is somehow compensated, it is extremely difficult to imagine how anything could be imbued with lasting significance.[8]

The religious visions of mankind have usually pointed toward something or someone that saves the world of nature and of human experience from vanishing into a total nothingness. Religions have, of course, been deeply affected by the passing of things. But they have sensed behind the facade of transience something that endures and, in enduring, preserves the past moments of our personal experience and of the universe's becoming from utter oblivion. Religion, Whitehead says,

> . . .is the vision of something which stands beyond, behind, and within, the passing flux of immediate things; something which is real, and yet waiting to be realized; something which is a remote possibility, and yet the greatest of present facts; something that gives meaning to all that passes and yet eludes apprehension; something whose possession is the final good, and yet is beyond all reach; something that is the ultimate ideal, and the hopeless quest.[9]

Such a vision responds to our desire not to be utterly forgotten. But can we honestly reconcile it with the fact of perishing that runs throughout the physical universe?

As William James emphasized, it is hardly possible to harmonize the religious vision of "something that gives meaning to all that passes, and yet eludes apprehension," with a materialistic cosmology. The bits of matter in the classical picture of physical reality are incapable of being taken up into any sensitive salvific scheme. Since they are by nature insensitive, they cannot be endowed with any significance other than mathematical. Only a universe in which "qualitative" (aesthetically valuable) experience is itself fundamental, is compatible with an interpretation of reality in which nothing is absolutely lost and every event somehow makes a lasting impact on the universe. Thus it is not a matter of indifference whether we envision the basic constituents of physical reality as insensate lumps of matter or as percipient occasions.

It is of course also true that religious and philosophical reflection have at times failed to take seriously the fact of perishing, consigning it to the realm of "mere appearance." And theologies have also at times superficially reified the intuited divine source of permanence and distanced it from all contact with becoming. Whenever this has occurred the religious vision has suffered either from escapism or a sense of deadness. But at least at other times religious consciousness seems to have been searching for a reality that embraces the flow of perishing events without eliminating the fact of perishing, and that intimately experiences becoming and perishing without itself dissolving in the flux.[10] Religious consciousness has symbolically groped for a transcendent reality that salvages from the flow of becoming and perishing events whatever of value has appeared in the course of the world's movement in time. And quite often sensitive persons and communities have been convinced that such a reality has manifested itself to them decisively by some mode of "revelation."

How, though, more specifically, is this religious vision of a permanence embracing the stream of perishing events capable of being harmonized with a coherent view of physical reality? Precisely how can occasions that have perished and things of value that have been pushed into an apparently irrecoverable past still be felt in the present? How, in other words, might we be able to join our hope that they have abiding significance with the fact of nature's becoming and impermanence?

In response to this set of questions we must again call to mind the notions of physical reality, perception and causation described in the earlier chapters, and the aesthetic notion of value sketched in the previous one. Intrinsic to this combination of concepts there is a framework within which we may legitimately see the transience of events and of valued entities as redeemable from absolute loss and as, therefore, capable of making a difference of value to the universe as we experience it in the present.

The structure of actual occasions as perceptive, as actively inheriting and synthesizing the past, and leaving themselves open to synthesis by subsequent occasions, gives us a basis for affirming the continued influence of occasions on subsequent occasions in the cosmic process. Every occasion is at root experiential, and the material of its experience is, at least in a vague sense, the entire universe. In perishing, occasions are not consigned to total nothingness but are granted a kind of "immortality" as elements in the experiences of subsequent occasions and groupings of occasions.[11] The aesthetic intensity of feeling that they enjoyed is deposited in the stream of events, so that when the immediacy of their own experiencing is past they continue to "survive" in varying degrees of vividness or dimness in the feelings of later entities. It is precisely in perishing that they deliver themselves over to assimilation by subsequent phases of cosmic process. It is their perishing that allows them to enter internally into others as abiding causal influences on the course of events. And this emptying of themselves, this non-clinging to their own immediacy gives them therewith a status of permanence in the experiential universe. Consequently, there can be no absolute loss in the universe since each perishing occasion is somehow felt by the perceptivity that is intrinsic to all the constituents of nature. It is because occasions are by nature transitional that there can be such a reality as causal efficacy or transfer of influence in the universe. And it is also because occasions perish that there can be a constant opening for the incursion of novelty into the emergent process.

The fact of perishing, then, does not vitiate purpose. Perishing allows occasions to sacrifice the vividness and immediacy of present enjoyment in order that they might have a bearing on others. And even if their impact is felt only in a dim way they have still made their valued contribution; and that contribution reverberates throughout reality, permanently.

In terms of this cosmic fabric of transitory experiences and perishable societies of occasions we may come to understand an aspect of the idea of God that religious traditions seem to be unanimous in affirming but by no means always unambiguous in expressing. This is the idea of divine care. What would such an idea mean in terms of our cosmography?

God might be understood here as the ultimate recipient of all the experiences that make up the cosmic process. As such, God would retain in increasing richness of aesthetic feeling all the vividness of immediate enjoyment that characterizes each entity, even though the constituent occasions in perishing have themselves lost this sense of present vividness. Divine care would weave into itself all of the experiences, becoming, living, emergence, destruction, disorder, entropy, conflicts, sufferings and dyings that occur in nature. Thus God's experience salvages what is apparently lost in the transience of events. As Whitehead says, God ". . . saves the world as it passes into the immediacy of his own life."[12] God experiences every actuality in such a way that its becoming, experience and perishing "aesthetically" enrich the divine life. The world, with its becoming, emergence and its dying, matters to God. In God's care for it, therefore, it finds its purpose and achieves its aim toward beauty. Whitehead himself has put it as follows:

> The wisdom of [God's] subjective aim prehends every actuality for what it can be in such a perfected system--its sufferings, its sorrows, its failures, its triumphs, its immediacies of joy-- woven by rightness of feeling into the harmony of the universal feeling, which is always immediate, always many, always one, always with novel advance, moving onward and never perishing. The revolts of destructive evil, purely self-regarding are dismissed into their triviality of merely individual facts; and yet the good they did achieve in individual joy, in individual sorrow, in the introduction of needed contrast is yet saved by its relation to the completed whole. The image--and it is but an image--the image under which this operative growth of God's nature is best conceived, is that of a tender care that nothing be lost.[13]

John Cobb draws out the possible implications of this vision of divine care for our own individual quest for significance:

> . . . just as some fragments of the past are taken up vividly into our new human experiences, so all things in the world are taken up into God's experience. Whatever we do makes a difference to God. In that case, we cannot regard our

slightest acts as finally unimportant. Further, what is taken up into God is not primarily our public behavior; it is our experience in the full intimacy of its subjective immediacy. Our deepest thoughts and most private feelings matter to us because they matter to God.

. . .

Not only does God experience our experience and include it within his own, but also in him there is no transience or loss. The value that is attained is attained forever. In him passage and change can mean only growth. Apart from God, time is perpetual perishing. Because of him, the achievements of the world are cumulative. It is this aspect of the vision of God which ultimately sustains us in the assurance that life is worth living and that our experience matters ultimately.[14]

In this interpretation of divine care God both feels and is felt deeply by the experiential occasions that make up the emergent universe. Perhaps at the level of primary perception human as well as other kinds of occasions already have a causal "awareness" of this pervasive sensitivity and at least a vague sense of being deeply felt themselves. Religious symbols and myths of divine care may be enigmatic expressions of this primordial perception. The word "God" and stories of God could then be understood as attempts to focus more sharply and linguistically the opaque but powerful feeling given in primary perception that something permanent and preservative runs through the becoming and perishing of events. Our reference to such a deeper dimension may require that we already resonate with some culturally contingent set of symbols of divine care and cosmic love. Such symbols may for some of us at least partially express what we feel obscurely in our primary perception as a universal purposiveness. At the pole of primary perception, according to hints given by art, poetry and especially religious expression, we have a dim feeling of the entire universe as well as a feeling of being assimilated by the processive universe in its deeper emergent comprehension of lower dimensions. There is an inarticulate causal feeling of the moments of our own experience both synthesizing the whole and being synthesized into further emergent depths of the whole. We have suggested that the notion of God points to the element of everlastingness within this flow of mutual sensitivity in the universe.[15]

God is understood here both as being felt and as actively feeling. In the mode of being felt the divine presence in the universe is a lure or power of persuasion continually offering new possibilities of intensity and harmony to patterns of events in cosmic becoming.[16] "God" is a word we may use to refer to the radical source of novelty and order in the emergent universe.[17] Without such a source of novelty and order we may well ask whether there would be any emergence at all. We may even question whether there would be a universe. For in order for anything to be actual it would have to be ordered or patterned in some way or other.[18] As we shall see in the next chapter it is a dubious theological or cosmological procedure to associate God too closely with the fact of order. For in addition to instances of order in our universe there are also the fact of novelty and the disturbing, disruptive effects that novelty may have on prevailing patterns of order. Nonetheless, it does not seem out of line to associate the religious sense of the divine with a metaphysically required ground of order in the cosmos, provided that we also associate God with the fact of novelty. Thus God, as principle of order and novelty, is understood here as the creative ground of the emergent universe, giving it its very being and presenting it with whatever possibilities for further emergent self-transcendence may be relevant at each particular phase.

At the same time, in this scheme of interpretation, God is always actively feeling all aspects of the universe in an unsurpassably intimate way. Since the occasions that make up the universe are themselves through and through experiential by nature, there is nothing incongruous in our holding that the creative ground of the universe is also the ultimate experiencing recipient of the events of world process. By virtue of the reception of these events into divine experience they are saved from any absolute perishing. Attributing this experiential-receptive quality to God makes it at lease conceivable, without departing from a consistent cosmology, to appreciate Tillich's summation of the religious response to the anxiety of being forgotten:

> Nothing truly real is forgotten eternally, because everything real comes from eternity and goes to eternity. And I speak now of all individual men and not solely of man. Nothing in the universe is unknown, nothing real is ultimately forgotten. The atom that moves in an immeasurable path today and the atom that moved in an immeasurable path billions of years ago are rooted in the eternal ground. There is no absolute, no completely forgotten past, because the past, like the future is rooted in the divine life. Nothing is completely pushed into the past. Nothing real is absolutely lost and forgotten. We are together with everything real in the divine life.[19]

Conclusion

Religious symbolism pointing to some ultimate context of cosmic significance, to a ground of meaning and love, to a comprehensive preservative care, is at least not incompatible with what we now know about the logic of emergence and the nature of physical reality. But can we go even further and maintain that the universe, if it is to become intelligible to us today, actually <u>requires</u> a religious interpretation? Whitehead, along with many other important thinkers, has insisted that it does. However, it seems to me that a necessary condition for ascribing to such a view is that we would already have an explicit trust in the universe as rooted in a "tender care that nothing be lost." We would not know what religious symbols were themselves pointing to did we not already have a primal sense of confidence in the ultimate goodness and meaningfulness of reality. But often this latent trust does not come to the surface until it is expressed mythically or religiously. And one does not generally encounter such forms of symbolism except in the context of a community of shared meaning and hope. Consequently, any possibility of interpreting the universe religiously might also entail that one already participates in and is committed to the beliefs of a community of faith. Could it be though, that such communities of faith carry in themselves the world's emergent impulse toward being comprehended by deeper dimensions of harmonized intensity and aesthetic enrichment? In any case I doubt if a sense of the world's general aim toward value can be deeply felt by those who have not experienced the urge to participate in a community of faith, where faith is understood as an adventurous openness and exploratory hope. For us humans it may well be that the quest for cosmic purpose coincides with the search for this kind of a faith community.

Chapter VII

ADVENTURE

The notions of physical reality and emergence that we have advanced in the previous chapters, unlike those of scientific materialism and mechanism, do not preclude in principle our attributing a teleological aspect to the universe. In fact they appear to be quite consistent with a religious affirmation of divine care at the heart of cosmic occurrence. However, we may have been too precipitous in offering this contention. Perhaps for the sake of taking things step by step in the interest of clarity, we have had to suspend momentarily explicit consideration of some complicating issues along the way. Our argument that nature is not incompatible with a religious interpretation may appear at times to have been too neglectful of questions that would seriously challenge a teleological view were we to consider them more extensively. Although the present chapter cannot compensate completely for these failures, it will attempt at least to treat more explicitly the most serious problem of all--the so-called problem of theodicy.

Still demanding our attention is the question of how to reconcile belief in an ultimate "ordering principle" conceivable in terms of the logic of emergence, with the obvious fact of disorder in the universe. What, more precisely, would be the nature of such a principle of emergent modes of order? What sort of power or capacity to influence would pertain to this transcendent dimension? Would not the fact of transience that besets all harmonious, intense instances of order (life, mind, civilization, for example) entail an inability on God's part to order the universe? And would not this limitedness mean that we would be compelled logically to reject the idea of an all-powerful God?

These and like questions are usually referred to as the theodicy problem. If God is all-good (perfectly loving and caring) and all-powerful, then disorder would not be allowed to appear in the universe. But disorder is manifestly present. Therefore, how can God's existence be morally or rationally affirmed? Thus goes the traditional formulation of the problem. In this chapter I shall briefly sketch the problem of God and evil in terms of concepts developed earlier, and I shall propose that one important way to deal with the theodicy problem is to associate the idea of God not only with order but also with <u>adventure</u>.

It is generally expected by all those who have any familiarity with philosophical and theological reflections on the problem of evil that no rationally or emotionally satisfying response to the fact of disorder will ever appear on paper. Merely speculative attempts always fall miserably short of reaching anything like a solution. Nonetheless, the overwhelming presence of chaotic elements in our experience simply demands that we ponder the matter. We cannot avoid speculating even though we realize that speculating does not solve the problem of suffering, death and all the tragedies that constitute what we call evil. Speculation, however, impotent as it is in relieving individual suffering, may still help us theoretically reconcile human symbols of hope and trust with the physical universe out of which disorder arises. Such theorizing is inadequate, but it is also indispensable.

We must begin by casting suspicion on those speculative theodicies that postulate a facile divine harmony or rational world order as the solution to the problem of evil. Rigid teleologies that impose a trivial form of harmony upon the universe and make God the overseer of this world order are intolerable not only on scientific but also on religious grounds. Nothing is more alien to an authentically religious outlook or to a belief in the unique value of individual personality than is a simplistic belief in world order. The Russian philospher Nicolai Berydaev vigorously chastises all theologies that attempt to force premature teleological interpretations on the universe:

> What value does the very idea of world order, world harmony possess, and could it ever in the least justify the unjust suffering of personality?[1]
>
> . . .
>
> World harmony is a false and an enslaving idea. One must get free of it for the sake of the dignity of personality.[2]
>
> . . . God is not world providence, that is to say not a ruler and sovereign of the universe, not <u>pantocrator.</u> God is freedom and meaning, love and sacrifice. . . The good news of the approach of the Kingdom of God is set in opposition to the world order. It means the end of false harmony which is founded upon the realm of the common. . . There is no need to justify, we have no right to justify, all the unhappiness, all the suffering and evil in the world with the help of the idea of God as Providence and Sovereign of the Universe.[3]
>
> . . .
>
> God is in the child which has shed tears, and not in the world order by which those tears are said to be justified.[4]

In Berdyaev's protest we have an intensely religious point of view that disassociates the human search for meaning from strict teleology. An obsession with overall harmony is not only unnecessary to the religious concern for personal significance; it is actually incompatible with it. A strict teleology is not required in order to affirm the radical worthwhileness of our lives in the scheme of events.

It is easy to sympathize with Berdyaev's strong opposition to teleologies that swallow up the individual and suppress personal uniqueness for the sake of an overarching concord. Any vision of the universe that dilutes the significance of individuals and their private suffering by sacrificing them to some abstract totality is repugnant. We do not need to review Kierkegaard's critique of Hegel to establish this point. And yet some form of teleology, some vision of cosmic purpose, is required precisely in order to make possible at all an appreciation of the unique value of persons and the poignancy of their experience of evil. It is absolutely imperative that we not abandon the quest for purpose in nature when we are concerned with validating the dignity of human personality. I am afraid that too much individualism and personalism have themselves acquiesced in a dualistic mythology that abandons nature to mechanistic desolateness. They have based themselves, as in the case of Berdyaev, on a radical dichotomy of nature and person, or of nature and history. When nature is understood after the fashion of scientific materialism this cleavage is quite intelligible For the only way to salvage the singularity of persons out of the deterministic commonality of a materialistic view of nature is to locate the core of personality at a separate level of reality over against the impersonal universe. But once we realize the abstractness and remoteness from experience of a materialistic cosmology, we must also question the validity of attempts to define personality or to ground its dignity apart from nature. There is no longer any need to disengage the exaltation of personality from the quest for cosmic purpose.

Instead it is appropriate both scientifically and religiously, that in our reflections on the problem of evil and personal suffering we examine further the implications of what we have called a loose teleology. For only according to some vision of cosmic purpose is it possible to establish the worth of the individual in the face of the negativities of his life. Acknowledging the "backing of the universe" is a prerequisite of any deep personalism. Without a conviction about the universe's general capacity to sustain significance there is insufficient basis for our affirming the incomparable importance of the individual.

The idea of a purposeful universe can be redeemed from the totalitarian rigidity that Berdyaev fears if it is allied to the notion of adventure. Adventure is the ingredient that will "loosen" teleology so as to give vitality to purpose and freshness to harmony. Without an aim toward order, movement in the direction of novelty would make the universe drift toward complete chaos. But without an adventurous advance toward novelty and freshness the universe would be frozen into utter sameness.

Adventure is the universe's search for continually more intense forms of ordered novelty.[5] If the actual world is a process, composed of becoming and perishing occasions then its movement toward integrating these occasions into ever richer modes of order may be called adventure. The term adventure always implies risk, the possibility of tragic loss, of failure to achieve the desired perfection of harmony. However, without adventure reality at any level of emergence lapses into decadence. There can be no standing still in a processive universe. Things must follow either the course of entropy or of adventure. "Pure conservatism" would be a violation of the very essence of the universe.[6] Thus if we are to speak of a cosmic aim or of nature's purpose we must recognize that its realization would occur only along the exciting but treacherous pathway of adventure. A kind of restlessness is intrinsic to all phases of emergence, a cosmic discontent, a sense that further ideals of harmonized intensity yet remain to be realized. Adventure is the hazardous undertaking of the quest for these novel ideals.

The notion of God, if it is to correspond to the essence of the universe, must be allied with that of adventure. Unfortunately, such an alliance has not always (perhaps not even often) prevailed in the religious life and thought of theists. The notion of God has been too closely associated with order at the expense of novelty and adventure.[7]

Associating God exclusively with order, a constant temptation of religion, leads to the decay of religious life and thought. Of course it is not possible to think of God apart from the fact of order since some kind of ordering is essential to the very being of entities. But traditionally theology has tended predominantly toward associating the idea of God almost exclusively with cosmic order and has been unaware of, or else has ignored, the fact of creative advance in the universe. Thus when the classical picture of world harmony broke down after being eroded by scientific theories of evolution and entropy, its naively conceived divine counterpart also vanished--quite fortunately we may say in retrospect. The biblical insight that God could also be understood in terms of novelty and adventure, risk and suffering had been suppressed; and so theology usually left out the question of how to relate the divine to the incursion of freshness into the world. Instead it focused narrowly on the question of how to relate God to the fact of order. And, as a result, the theodicy problem was tragically and erroneously formulated. For in addition to order there is also the fact of novelty. And, further, there is the fact of inevitable disruption of established patterns of order wherever there is an influx of novelty. When we speak of God's relation to nature, we must do so in terms not only of actualized order but also of the universe's adventure toward novel forms of order. This dual reference will provide us with a richer context for locating the fact of evil.

We have emphasized throughout this work that our universe is a process and that this process is characterized by a creative advance from lower toward higher emergent dimensions. Therefore, in attempting to relate the notion of divine causation to such a universe we have to understand God not only as the ground of whatever order happens to have emerged, but also as the creative pull that energizes the world's ever becoming "more". In persuading the cosmos toward more complex and intense modes of emergence, however, this transcendent influence leaves itself exposed to the charge that it is responsible for the discord that inevitably accompanies an ingression of novelty into any particular situation of order. In the advance toward new patterns of complexity there is always a risk that previously realized order will be destroyed, or that novel complexity will overwhelm established harmonies. Therefore, whatever source of power motivates the world process toward emergent novelty would appear to be the ultimate reason that there is the evil of disorder in the world. Does this mean, then, that God is responsible for evil? And, if so, does the world's advance toward novelty justify the presence of evil?

In order to approach this troubling question we should first examine more carefully what is meant by evil. Evil is not simply identifiable with perishing or with chaos. It may also be associated with unnecessary, out-of-season triviality. Any situation of destruction and disorder is evil. But so also are those harmonious, undisturbed situations where a richer wholeness is attainable and yet there persists an obsession with partiality. "There is then the evil of triviality--a sketch in place of a full picture."[8] A suppression of the universe's perpetual urge toward adventure is a turning away from the value of aesthetic intensity. Clinging to low-grade forms of harmony unnecessarily is a deviation from the good. To remain content with monotony when further variation is relevant and possible is infidelity to the cosmos.

Thus when we use the term "evil" we are indicating not only instances of physical and moral disintegration. We are also referring to situations where there may indeed be a stable harmony and order but also an absence of zest for intensity, an uncalled-for lack of adventure and a fear of novelty. The untimely refusal to experiment with newness may in certain circumstances also count as evil. This means that the refusal to hazard the possible discord that threatens every creative advance may constitute evil just as much as does a circumstance involving outright disharmony.

In the light of this broader description of evil we should reformulate the theodicy problem so as to ask not only about the justification of disorder in a world created by an allegedly all-good and all-powerful God, but also about a world that seemingly cannot exist apart from an intrinsic adventurousness. Is a God who stimulates the world toward creative advance by offering it ever new possibilities morally justifiable, given that the incursion of novelty brings with it the risk of chaos?[9]

The only way to begin a response to this question is to consider the alternative. Could there conceivably exist a divinely created universe other than one endowed with a potential to evolve, and in evolving, to risk the constant threat of disorder? Is any other universe possible, metaphysically speaking? Try to conceive of an absolutely static world, totally and perfectly ordered by God in every possible way. Such a universe is not even conceivable. For a totally immobile and completely ordered world would not be distinguishable from its creator and, therefore, would not be a world. A world without internal, self-initiated movement, without any possibility of deviation from a divinely imposed scheme of order, would be a sheer emanation or extension of its maker's own being. It would have no autonomy, integrity or self-coherence. That is, it would not exist as a distinct reality which God could transcend in any way. It is impossible metaphysically for a creator to create a non-becoming, perfectly ordered world. The idea of a perfect universe is a contradiction in terms. The only conceivable world that would be compatible with the notion of God is one in which there is a possibility of creative advance, of adventure with its inevitable risk of discord. Thus, if God is ultimately "responsible" for the disorder attendant to the cosmic adventure, this responsiblity cannot be equated with reprehensibility.[10]

Still, recognizing the theoretical congruence of an adventurous emergent universe with the notion of a transcendent principle of order and novelty hardly solves the theodicy problem. Religious experience and expressions (I am thinking particularly, but not exclusively of biblically based religious types) profess belief in a God of love and care. In terms of these religious convictions the problem of theodicy is also fundamentally that of how to harmonize the pervasive fact of disintegration and chaos that accompany adventure, with the alleged love and concern of God for the universe.

Obviously this aspect of the theodicy problem calls for a clarification of the meaning of words like love, care and concern. But here we run into immediate difficulties. Conceivably, one might suspect, we could define love (and, therefore, "God") in such a sweeping and flexible manner as to make it compatible with each and every imaginable situation. As a matter of fact some contemporary critics of theism have accused theists of making God's existence compatible with any conceivable situation of cosmic or personal disorder. Then it would be impossible to show what difference divine love might make to us. And if we cannot show what difference it would make, what reason is there to believe in its presence in the universe? God's existence would be "unfalsifiable" and, therefore, outside the sphere of meaningful reference. If there is nothing that could ever conceivably count against God's love, then there is little sense in our putting our trust in this love.[11]

This critique is important because it compels us to scrutinize more carefully the possible nature of divine love. It is a healthy protest against a shallow theism. However, I do not think that in the final analysis the critique is fully justifiable. For there is, after all, as the critics themselves point out, a situation conceivable that would count against the existence of a loving God. This would be a situation in which "God" were able to eliminate suffering and evil and yet refused to do so. Such a circumstance would indeed force us to abandon theism in the very name of love.

In our definition of love we are inevitably influenced and constrained by our own interpersonal experience. And we know from this experience that love is compatible with some circumstances and incompatible with others. For example, when our projects are subverted by the jealousy and hostility of others, we realize that this is inconsistent with love. Or when we find that others are able to assist us when we are in pain or need, and yet they needlessly refuse to do so, we interpret this apathy as incompatible with love. However, a situation in which others are willing but <u>not able</u> to offer us immediate deliverance from suffering is certainly consistent with their love for us. We can all think of instances in our own experience where the helplessness of friends and loved ones to extricate us from distress detracted in no way from our feeling that they cared deeply for us. Even in their weakness, as it turns out, they communicated to us a sense of strength and courage that would not otherwise have been available to us. In fact their inability to assist us in an immediate way may have left an opening for the welling up in us of a more enduring sense of potency at a deeper level of our being.

It seems to me that any meaningful theodicy should employ the analogy of this loving helplessness when it reflects on God and God's relation to a world in which disharmony is a recurrent fact. Consigned as we are to utilizing elements of our own experience in our symbolization of ultimacy, we should at least attempt to base our models of deity in those experiences that move us most deeply, that allow us to grow in interior strength, and that give us a sense of being deeply cared for in spite of fateful threats to our existence. Such experiences include not only those in which others directly help us, but also those in which they express concern for us, even though they may be helpless to deliver us.

May God be conceived of in terms of such a model of loving helplessness? If we are to talk of God in an adult way today, must we not at least experiment with such a notion? It is thoroughly repugnant to our own sense of compassion to hold that God actually has the power directly to intervene in human suffering, or in any situation of moral or physical evil, and yet refrains from utilizing this power. Such a posture would simply not be compatible with love in any humanly meaningful sense of the term. Such a God, as Camus has protested with convincing arguments, could hardly arouse us to a sense of compassion for others.[12] How could a loving God have permitted the monstrosities of Auschwitz, Bangladesh, or Cambodia, the tortures of innocents throughout the ages, and the incredible loss of intense harmonies in the adventure of cosmic emergence? Is it totally out of the question for the believer to respond in full seriousness that <u>God could not help it?</u>

Of course such a response is seemingly replete with serious problems. Above all: in what sense might we legitimately attribute "helplessness" to an allegedly all-powerful God? Further: what is the meaning of "power"? And is helplessness always incompatible with perfect power?

Power and Adventure

Could God have prevented Auschwitz? Could God reverse the flow of entropy? Is God able to create a universe without having it emerge through the turmoil of an evolutionary process? Could not the species that inhabit the earth have come about without all the struggle and loss that evolutionary theory portrays? Need chance have played such an important role in evolution if God is in any way the principle of cosmic order? In short, could not God have prevented the incredible anguish, suffering and waste that ravish the earth?

In whatever way the theodicy problem is formulated, there is usually a hidden premise about the meaning of "power", "potency", "ability", "can", "could", etc. We need to explore the latent assumptions about the meaning of these terms when we ask whether God had or has the "power" to eliminate evil and prevent discord.

Fundamentally power means the capacity to influence. In other words, power implies the potential for causal efficacy. As we noted in Chapter III, however, causation has often been narrowly construed in accordance with now unacceptable concepts of physical reality and perception. Unfortunately, when God's capacity to influence has been theologically expressed, it has often been represented in terms of incoherent ideas of causation, perception and physical reality. As such, divine power has been imagined as the coercive, overwhelming, forceful impact of one (active) entity on the (passive) receptivity of a second. That is to say, the idea of divine causation has been shaped in accordance with our experience at the level of seconday perception, and in fidelity to the dominant view that this kind of perception is fundamental.

However, by understanding the notion of causation as an instance of the more basic fact of _primary_ perception, we have moved away from the conventional materialistic, push-and-shove, view of causal efficacy. We should now transfer the results of this reinterpretation to the notion of divine influence as well. In doing so we may be able to reconcile the idea of God's power with that of a loving helplessness vis-a-vis the fact of evil in the cosmic adventure.

What makes it possible for entities to influence one another in a fundamental way is the fact of perceptivity (prehensiveness) that defines each actual occasion. Each occasion becomes itself by <u>actively</u>, synthetically prehending the data that have been presented to it by the perishing of previous occasions in a series. Because each occasion actively appropriates its past (in a manner determined also by the way in which it receives novel possibilities into its experience), it is, in a sense, self-caused, a <u>causa sui</u>.[13] There is no absolute passivity in this fundamental mode of perception and causation. Each occasion is, to some degree, actively perceptive, actively inheriting and integrating the "objective" past, actively screening, embracing, refusing and "deciding" upon what novel elements shall make up its own unique feeling, enjoyment and satisfaction. Although in the case of aggregates of occasions available to secondary perception causal impact appears largely in the form of active agents moving passive recipients (as in the impact of one billiard ball upon another), in the microcosmic realm of occasions of experience causal influence involves an intensely active receptivity to previous events and future possibilities.

Therefore, if nature is in any way affected by God at a <u>fundamental</u> level, its being influenced would not have to be imagined as a completely passive reception of divine causal force. If nature were totally pliable in the hands of a coercive "ordering principle" it would be impossible for us to make a clear conceptual distinction between God and the world. Religious experience, however, demands that we make such a distinction (which is not the same thing as a separation), for otherwise we would jeopardize the transcendence of the divine. And cosmological reflection also requires that we attribute a certain indeterminacy to world process, a quasi-resistance on the part of the cosmos to any hypothetical divine activity. For without some aspect of intransigency before God, the world would not really be a world. If it were absolutely malleable it would lose all distinctness from its "orderer". Residing in the self-causative aspect of each occasion of experience, however, there is a certain "freedom" or indeterminacy at a fundamental level of the world's being. Only some such self-determining potential insures that the world does not dissolve immediately into God's own being.

Given the perceptive, experiential-mental basis in physical reality for a coherent doctrine of causation, we can now conceive of the divine capacity to influence the cosmos in terms consistent with a divine "helplessness" to remedy concrete instances of disorder in the universe. Moreover, we may do so in a manner quite in line with a genuinely adult notion of love as well.

When one entity causally influences another at the fundamental level of cosmic occurrence, it does so by perishing and handing itself over as an objective datum to be appropriated in an indeterminate way by subsequent occasions. It does not coerce the immediately subsequent occasion to conform to its own structure or quality of feeling in any absolutely rigid fashion. For each occasion actively inherits its past in a unique way. By simultaneously prehending (through its "mental pole") possibilities for advancing beyond or simply carrying on the specific tone of feeling transferred to it by previous occasions, each occasion asserts its self-causative singularity. So there is no coerciveness, but rather a kind of persuasiveness that characterizes this primary causation.

Divine causation would not be an exception to but rather an exemplification of the non-coercive character of influence that is required by occasions at the base of the world process. In order to conceive of divine causation we should not take as our point of departure the crude images of transfer of power that we find in the objects of secondary (sense) perception. Instead we should look to the finer transactions that occur in primary perception (in the occasions of our own as well as all cosmic experience). At the primary pole of perception and causation coerciveness is ruled out, and persuasiveness is the dominant mode of influence. We must not project into our notions of divine causation the commandeering forcefulness that seems to rule the relations among aggregates observed by secondary perception (rocks colliding with one another, hands moving objects, potters molding clay, etc.) Again, it is the unwarranted belief that sense-perception is fundamental that has led to such over-simplified, abstract ideas concerning causation. Traditional forms of theism, having applied these derivative ideas to God, have been led toward untenable conceptions of divine power. And as a result the theodicy problem has been misleadingly formulated.

God's power is persuasive rather than coercive.[14] What we earlier referred to as divine "helplessness" may now be understood in part at least as non-coerciveness. Divine love and power may both be interpreted in terms of the notion of persuasiveness. We detract in no way from the sublimity of divine power when we label it persuasive rather than coercive. For persuasive power, in the fundamental sphere of natural occurrence, is inexpressibly more capable of exercising influence than coercive power would be. God's capacity to influence can be exercised more radically and internally on the cosmos if it is non-coercive than if it were a compulsive transfer of energy like that which we discern among the relatively abstract entities at the secondary pole of perception. In the latter, derivative kind of causation that pervades the massive material objects studied by classical physics, transfer of power is purely external. It does not penetrate to the heart of its effects. In primary perception, however, the cause is "freely" internalized by the prehending occasion. And so the cause endures as an abiding constituent <u>within</u> the effect precisely because it does not coerce from outside. Classical theories of causation are unable to explain how a cause can have such an abiding influence on its effects.

As the occasions of the world-process (including those of our own experience) appropriate the power of a divine ordering principle and ground of novelty at the primary pole of perception, they are not forced into a deterministic response to God's influence. There remains a certain play for self-causation and self-transcendence in percipient occasions. God's influence is such that it allows the occasions to be themselves. It does not overwhelm or force them into more intense forms of relationships. God does not compel the universe toward further adventure. Divine love gently persuades the world of occasions toward the realization of further relevant possibilities and intensity of enjoyment; but in doing so it allows scope for deviation from the patterns of ever richer harmony and intensity that are held out to it as possibilities. Divine power is helpless to prevent the cosmic process from turning aside occasionally into chaos or from remaining stuck in banality. The movement of the universe toward increasingly intense and complex configurations of physical reality or moral order does not guarantee that tragedy or periods of stagnation will not accompany the adventure. Whether the adventure is worth the price, however, can be determined only according to criteria that flow from an adult understanding of the nature of love.

Charles Birch expresses the view that the apparent helplessness of God to prevent earthquakes, peculiar genetic mutations, human horrors or the wasteful and random aspects of evolution in no way entails powerlessness except in terms of the crudest conceptions of power:

> Is God then powerless? No--there is power in love! There is power in persuasive love that is greater than all other sorts of power. There is no need for any other sort of power. It is because we are unconvinced of the power

of persuasive love that we want to invest God with dictatorial coercive power.[15]

Whitehead himself observed how our Western theological traditions have typically modelled their notions of divine causal efficacy on the image of rulers and despots, that is, on those whose power is coercive rather than persuasive: "The Church gave unto God the attributes which belonged exclusively to Caesar."[16] Instead our images of God should be shaped by our dwelling on ". . . the tender elements in the world, which slowly and in quietness operate by love. . ."[17]

Since our God-images are so often fashioned almost exclusively on the pattern of imperial power, we tangle ourselves in endless theological knots trying to excuse "providence" for failure to implement its alleged capacity to force order onto the cosmos. Yet if our images of God were to assume a form corresponding to our experience of the "tender elements in the world," we would be able to envision God's radical causal efficacy in a manner consistent with both perfect power and infinite love.

It is our fear of adventure and our obsession with safety that give rise to the projection of God as coercive power. Our desire to be magically extricated from all situations of disorder that inevitably accompany adventure precipitates numerous forms of self-abasement and hostility in the face of the imagined omnipotence of a divine potentate. It may never occur to us that perfect power is "made manifest in weakness,"[18] and that God's power is most efficacious in its "letting be," in its refraining from imposing a fixed harmony onto the world. A God of persuasive love, Birch says, ". . . does not guarantee the 'safety' of any creature in the universe. Love never does. Mature religion can accept this."[19]

Mature religion, more than anything else, is characterized by the "high hope of adventure"[20] that has its roots in the cosmos and its creative ground. Instead of being obsessed with maintaining established forms of order, it shares in and carries on the emergent universe's experimentation with novelty. As such, mature religion is willing to risk the threat of discord that attends any evolutionary zest for realizing more intense forms of harmony at the opportune moments. A religious spirit that combines a delight in beauty with a taste for adventure lives harmoniously with a universe in which perishing, tragedy and freshness of patterned intensity are constant elements. And when it speaks of God it refers to a principle of novelty as well as of order, of adventure along with peace.

Unfortunately, what goes by the name religion has often sided with triviality. Religion has not seldom succumbed to the evil of monotony for the sake of avoiding the risk of adventure.[21] This religious "anaesthesia" comes to expression particularly in those theodicies that talk of God only in terms of cosmic order and that neglect the question of novelty. It is such devotion to false harmony that Berdyaev legitimately excoriates. However, the perspective on theodicy adopted here is one in which the evil of chaos, monstrous as it is, does not negate the value, significance or purpose that resides in a universe of adventure and tragic beauty. In a full sense of the term, "religion" would mean, therefore, an attitude of accepting reverently and hopefully the essence of the universe as a divine adventure toward novel, more intense forms of order. In this sense it would be representative of what we earlier called faith. An adventurous religious faith would suffer along with the cosmos as it struggles precariously and at times tragically toward realization of the beauty that constitutes cosmic purpose. The courage to accept the restlessness and loss in the cosmic adventure is given in a genuinely religious faith that all achievement of intensity of feeling as well as all perishing is finally salvaged and creatively transformed by God's own experience.

Emergence and Divine Tragedy

In the light of our notion of God as persuasive love, we may now also grasp why the principle of emergent order about which we speculated previously does not intrude into or violate the operations of lower emergent dimensions. Its fundamental essence is that of a loving "letting be." Its mode of influence is persuasion, and its creative style is that of proposing relevant possibilities of deeper aesthetic enjoyment but not forcing them upon the cosmic process. It offers ever newer and richer "extraneous" ordering principles, but it does not magically manipulate the workings of each successive dimension of cosmic emergence. It allows a progressively deeper integration of entities into richer modes of harmony, but it does not impose its organizational power. (This may help account for the chance and indeterminacy in physical and biological phenomena). Rather it invites, attracts, lures the cosmos toward novelty of patterned intensity, and is, therefore, the ground of evolutionary adventure and cosmic beauty. It preserves in its own experience all that from our limited perspective appears to be tragic loss in the process of emergence. But it does not need to suspend the laws and activities characteristic of the physical, chemical, biological, psychological and interpersonal dimensions in order to achieve its integration of the multiplicity of cosmic occasions into the unity of the divine life.

Conceived of on the model of tenderness rather than coercion, God "dwells in" and "relies upon" the workings of lower dimensions of cosmic emergence in order to realize the divine adventure toward intensity of feeling and enjoyment of beauty. An analogy may help to clarify this point: mental activity in some very loose sense may be compared to divine activity. Our own mental operations, to use Polanyi's terminology, dwell in and tacitly rely upon "subsidiary" biological and physico-chemical processes without in any way being reducible to, or explicable in terms of an analysis of, these subsidiaries. Our mental achievements cannot be fully explained by biologists, physicists or chemists; but still we must admit that these achievements could not occur without the faithful and predictable recurrence of physical and biological reactions and routines. Similarly God is not reducible to or explicable in terms of any lower level of cosmic emergence. God, as the ground of order and novelty, must be conceived of as a reality distinct from (though not separate from) the world. But may we not hold that a God of persuasive love, in a fashion analogous to our mental life, dwells within and even relies upon subsidiary cosmic processes as a condition for "divine" achievement? And would not the breakdown of any level of cosmic emergence, especially that of human personal interaction, result in the "failure" of the divine to implement its aim?

Such conjecturing has to be very tentative of course. But the analogy of divine creative achievement to human mental achievement seems to be harmonious both with the logic of emergence and with numerous religious symbols of God's power and love. Our analogy seems particularly congruous with the moving images of God's tragic vulnerability, embodiedness, suffering and even dying, of divine self-emptying or self-withdrawal that our religious traditions, occasionally at least, communicate to us.

Let us try to unfold this comparison a bit further. Our mental life, significant and valued as it is, is extremely vulnerable in the sense that its achievements are utterly dependent upon, though not fully explicable in terms of, the proper chemical reactions that occur in the cells of the brain and nervous system. As we know from the phenomena of senility, mental retardation, and other afflictions of the brain, the human mind is only partially actualized wherever there is cellular or organic impairment of the physiological base upon which it relies. If we could but restore or bring about the normal functioning of brain cells and nerve tissues, as well as their own subsidiary chemical activity in such cases, then mental life would come flooding back in fullness and richness. But without the proper preparation of its physical base, mental activity has nowhere in which to dwell. And so it fails to become incarnate.

This dependency or vulnerability of higher emergent activity with respect to lower seems to intensify as we move up the scale of emergence. I would even suggest now that God's own life may in a certain sense be comparable to our mental life both in its dwelling in a proportionate physiological base as well as in its vulnerability to and "dependence" upon the adequate preparation of this base. Such a suggestion is not out of line either with the logic of emergence or with the testimony of significant strains of religious symbolism. Accordingly, the actualization of the divine presence in the universe would have a fragility and precariousness even more delicately intense than that of our own minds with respect to their physical subsidiaries. For a God who dwells in and relies upon the various strata of cosmic process would be "dependent" not only on the performance of sub-human emergent dimensions, but also on the considerably less predictable interpersonal processes of human self-integration (or comparable "conscious" processes in other corners of the universe). The incarnate realization of divine life awaits the adequate preparation of its cosmic subsidiaries no less than the actualization of mental life requires the reliable institution of an extremely complex physico-chemical and biological substrate.

In the case of God's self-realization in the cosmic adventure there is a risk of tragedy proportionate to the delicacy of the subsidiaries that are integrated into the divine field of emergent possibilities. In the case of God these subsidiaries include personal centers endowed with a capacity not only to unite with one another on a planetary scale, but also to engage in mutual isolation and destruction. In a very real sense the birth of God, on our planet at least, awaits the outcome of our own human decision for entropy or emergence. And the question of the meaning of our lives may well be tied up with such a decision.

Conclusion

Creative advance takes place only along the borders of chaos.[22] In the world's transition from triviality toward aesthetic intensity there is the omnipresent risk of evil. "Evil is the half-way house between perfection and triviality."[23] The introduction of novelty into the world means that the past has to give way. As a consequence cosmic process inevitably involves perishing and discord.

In the turmoil of emergence, however, God's purpose is not that of precipitating chaos, but rather that of luring the universe toward heightened enjoyment and beauty. "God's purpose in the creative advance is the evocation of intensities."[24] And yet it must be admitted that in maximizing the qualitative aesthetic intensity of the cosmos its creative ground must be held responsible for at least some of the chaos that accompanies evolution.[25] Whether such a God is tolerable will probably depend in part at least on the degree to which adventure is considered important or necessary in one's life as well as in one's vision of reality. I would like to express my own agreement with John Cobb and David Griffin who hold that a God of adventure, while perhaps responsible for the evil of discord that accompanies novelty, is not indictable for it. Their statement summarizes what I have been leading up to in this chapter:

> God is partly responsible for most of what we normally call evil, i.e., the evil of discord. Had God not led the realm of finitude out of chaos into a cosmos that includes life, nothing

worthy of the term "suffering" would occur. Had God not lured the world on to the creation of beings with the capacity for conscious, rational self-determination, the distinctively human forms of evil on our planet would not occur. Hence God is responsible for these evils in the sense of having encouraged the world in the direction that made these evils possible. But unnecessary triviality is also evil, since it also detracts from the maximization of enjoyment. Hence, the question as to whether God is indictable for the world's evil reduces to the question as to whether the positive values enjoyed by the higher forms of actuality are worth the risk of the negative values, the sufferings.

. . .

Should we risk suffering, in order to have a shot at intense enjoyment? Or should we sacrifice intensity, in order to minimize possible grief? The divine reality, who not only enjoys all enjoyments but also suffers all sufferings, is an Adventurer, choosing the former mode, risking discord in the quest for the various types of perfection that are possible.[26]

CONCLUSION

As I have stated throughout this book, it is neither possible nor desirable to respond to the question of nature's purpose in a purely scientific way. Science is capable of dealing only with data that appear at or relate to the secondary pole of perception. At this pole "qualitative" dimensions of the cosmos (beauty, value, aim) have already been left behind for the most part, and what remains is readily transformed into a predominantly "quantitative" field of material for scientific analysis. Science is adequate for describing the relatively abstract remnants of our perceptive encounter with the cosmos. But we would be overburdening its limited methodological possibilities were we to expect it to set forth any statements concerning nature's purpose. Instead we must rely upon a kind of discourse that attempts to retrieve and express the data of primary perception. I have suggested that the language of religion may be understood as representing in a mythic and symbolic way at least a portion of the qualitative data given to us in primary perception.

The key to our grasping the relationship between science and religion lies especially in the notion of perception that we have developed in the preceding chapters (especially Ch. III). We have maintained that perception is an active process and not a simple passive reception of impressions. And we have also emphasized its polar nature: at the primary pole of perception we are sensitive to a vast array of cosmic qualities that are left behind by the lucid secondary projections of sense experience. Science, especially since the seventeenth century, has envisaged the world of its formal concern almost exclusively in terms of the quantifiable aspects of nature lifted out or abstracted by secondary (sense) perception. And, following this focus, cosmologists in the last three hundred years or so have usually ignored the "qualitative" value-laden data given in our primary perception. At times, even to this day as we have seen, they understand the beautiful, the valuable, the purposeful as mere projections of our own "subjective" desires and wishes back onto the blank indifference of the material objects abstracted by science. Their distrust of religious symbols, myths and teleological hypotheses is bound up with a failure to consider the possibility of a primary kind of perception in which beauty, value, significance and purpose constitute ontologically "real" data of experience rather than products of wishful thinking. The intellectual, cultural and theological confusion set loose by the incredibly narrow "modern" notion of perception is, to say the least, enormous.

Our view has been that both science and religion are rooted in experience but that each is based in a different region of the perceptive process. Religion, employing a mythic-symbolic language that is always culturally conditioned and historically contingent refers to the data of primary perception, while science seeks to express correlations among the objects sensed, abstracted or imagined at the pole of secondary perception. Both types of expression have reference to the real world but in quite different modes of symbolic reference. Scientific language, although often purporting to be fundamental, is actually descriptive of an already rather late and abstract realm of objects lending themselves to demarcation by the crisp logic of mathematics. Religious language, however, while lacking the precision of scientific-mathematical discourse, refers us as well to a more fundamental region of experience where the universe has not yet been sharply delineated by our senses or by a quantifying analysis. Just as science must constantly revise its models so as to surmount the deficiencies of its abstract (usually mathematical) models of nature, so also religions are called upon continually to revise their enigmatic representations of cosmic significance in keeping with primary perception's intuition of an ongoing cosmic adventure. Like science, religion is not always (perhaps not even often) inclined to heed this obligation of accommodating itself to the adventure of revision. As Whitehead accurately points out, much of the conflict between science and religion stems from the reluctance, especially on the part of religion, to embrace adventure. But the conflict also flows in great measure from scientific thought's failure to move beyond the "sensationalist" doctrine of perception. Perhaps the most impressive way in which scientific thought could display its own alleged espousal of the spirit of adventure today would be for it to review its motives for still clinging uncritically to a shallow notion of perception.

My objective in this book has not been one of demonstrating that there is purpose in nature or of specifying what that purpose is. Instead my intentions have been much more modest. I have simply tried to offer an alternative to the picture of nature inscribed so vividly in modern scientific thinking--a picture that in principle leaves no openings in nature's soil for the seeding of divine purpose. I cannot prove conclusively that the predominantly Whiteheadian cosmography I have sketched is adequate either. As a matter of fact I assume that it is not. Everything that I have written demands that we continually revise our representations of the universe, the present one not excepted. However, I do think that the picture of nature, perception, causation, and religion tentatively outlined in this book is more consistent not only with experience and logic but also with science itself than is the questionable materialism that still hovers over modern speculation on matter, life and consciousness.

While they are being uprooted in many quarters of modern thought the premises of scientific materialism and reductionist empiricism still have a tenacious hold on our consciousness. Even though philosophers like Polanyi and Whitehead have offered outstanding critiques of materialism, their thought has not yet penetrated deeply into our intellectual and cultural life. As a result we tend to carry around ridiculously outworn pictures of nature that resist not only teleological interpretations but even the insights of contemporary science. I hope that the present volume has made at least a small contribution to the important enterprise of changing our ideas of the universe so as to open it up more fully to the interpretations of contemporary science and adventurous religion.

FOOTNOTES

INTRODUCTION

[1] George Gaylord Simpson, The Meaning of Evolution, Revised edition. (New York: Bantam Books, 1971), pp. 314-15.

[2] Jacques Monod, Chance and Necessity, trans. by Austryn Wainhouse, (New York: Vintage Books, 1972) pp. 112-13.

[3] S. E. Luria, Life: The Unfinished Experiment (New York: Charles Scribner's Sons, 1973), p. 148. The statements of Simpson, Monod and Luria are reminiscent of some much earlier expressions of cosmic pessimism collected by John Hermann Randall in The Making of the Modern Mind (New York: Columbia University Press, 1976, pp. 577-621.

[4] Monod, for example, utilizes the principle of indeterminacy to support his vision of the biosphere emerging from pure chance: Chance and Necessity, pp. 114 ff.

[5] I am referring especially to the thought of Rudolf Bultmann, but also to other forms of existentialist theology that posit a Neo-Kantian dichotomy of nature and history. A major exception to this kind of theology is that of "process theology" based on insights of A. N. Whitehead and Charles Hartshorne. We shall draw upon much of this process theology later on in this book.

[6] Gordon Kaufmann, God the Problem (Cambridge: Harvard University Press, 1972), p. 122.

[7] Bertrand Russell, "A Free Man's Worship" in Mysticism and Logic (Garden City: Doubleday Anchor Books, 1957), p. 52.

[8] Theodore Roszak, in spite of his powers of thought and articulation, has not really

presented us with a solid philosophical base for his literate protests against dualism and objectivism. Cf. Where the Wasteland Ends (Garden City, New York: Doubleday Anchor Books, 1973). Nor does the recently popular work of E. F. Schumacher, A Guide for the Perplexed (New York: Harper and Row, 1979) offer a satisfactory base for a radical critique of scientism, in that, like other vitalistic reactions it concedes too much to mechanism.

9 Dean Turner's recent book is full of valuable insights on the issue of science and religion. But it fails to develop a consistent epistemological position and at times becomes excessively emotionalistic in its understandable frustrations with mechanism. Cf. Commitment to Care (Old Greenwich, Connecticut: Devin-Adair Company, 1978).

CHAPTER I: DUALISM

1 The term noosphere is used by Teilhard de Chardin to refer to the phase in evolution where consciousness as we attribute it to man becomes present and begins to spread over our planet. We shall propose with Whitehead and Hartshorne that while consciousness does not exist on earth prior to man, mentality is a pervasive aspect of physical reality.

2 Hans Jonas, The Phenomenon of Life (New York: Harper and Row, 1966), p. 9.

3 Ibid.

4 Ibid., pp. 9-10.

5 Ibid.

6 I am indebted in many ways to Jonas' book The Phenomenon of Life for this interpretation.

7 This expression is Paul Tillich's: Systematic Theology Vol. III (Chicago: University

of Chicago Press, 1963), p. 19.

[8] Cf. E. A. Burtt, <u>The Metaphysical Foundations of Modern Science</u> (Garden City, New York: Doubleday Anchor Books, 1954).

CHAPTER II: PHYSICAL REALITY

[1] R. G. Collingwood, <u>The Idea of Nature</u> (New York: Oxford University Press, 1960). p. 177.

[2] Alfred North Whitehead, <u>Modes of Thought</u> (New York: The Free Press, 1968), p. 156.

[3] Whitehead observes that "[the] strength of the theory of materialistic mechanism has been the demand, that no arbitrary breaks be introduced into nature, to eke out the collapse of an explanation. . . But if you start from the immediate facts of our psychological experience, as surely an empiricist should begin, you are at once led to the organic conception of nature. . ." <u>Science and the Modern World</u> (New York: The Free Press, 1967), p. 73. Whitehead derives the notion of "radical empiricism" from William James.

[4] Alfred North Whitehead, <u>Process and Reality</u>, Corrected Edition, ed., by David Ray Griffin and Donald W. Sherburne (New York: The Free Press, 1978), p. 190. Whitehead is employing an expression like that of William James: "Your acquaintance with reality grows literally by buds or drops of perception." (p. 68).

[5] The material in this section as well as in the book as a whole comes from my study of a number of works both by and about Whitehead. In particular Whitehead's own works <u>Process and Reality</u>, <u>Science and the Modern World</u>, <u>Adventures of Ideas</u>, <u>Religion in the Making</u> and <u>Modes of Thought</u> constitute the dominant background of the line of thought developed in this book. As I suggested earlier, however, I am attempting to

simplify Whitehead's thought considerably for the purpose of making it available to introductory readers.

6 The term "event," used predominantly in Whitehead's Science and the Modern World, does not have precisely the same meaning as "actual occasion" (the term used extensively in Process and Reality). However, for our purposes it is sufficient to indicate the "temporal-experiential" quality that both expressions are pointing to in the constituents of physical reality. In more precise language we should say that an "event" may actually be constituted by a number of "occasions." (Process and Reality, p. 73).

7 Whitehead, Modes of Thought, pp. 130-132.

8 Ibid., p. 136.

9 Cf. John Cobb, God and the World (Philadelphia: The Westminster Press, 1969), p. 70.

10 Whitehead, Modes of Thought, p. 138.

11 Ibid.

12 Quoted by Charles Birch, Nature and God. (Philadelphia: Westminster Press, 1965), p. 113.

13 Whitehead, Modes of Thought, p. 146: "For the modern view process, activity and change are the matters of fact. At an instant there is nothing. Each instant is only a way of grouping matters of fact. Thus since there are no instants, conceived as simple primary entities, there is no nature at an instant. Thus all the interrelations of matters of fact must involve transition in their esence. All realization involves implication in the creative advance."

14 Whitehead, Science and the Modern World, pp. 51, 58.

[15] Ibid.

[16] Cf. Whitehead, *Modes of Thought*, pp. 156-57.

CHAPTER III: PERCEPTION

[1] Monod, p. 30.

[2] Ibid., p. 180.

[3] Michael Polanyi and Harry Prosch, *Meaning* (Chicago: University of Chicago Press, 1975), p. 162.

[4] W. T. Stace, "Man Against Darkness," *The Atlantic Monthly* (Sept. 1948), p. 54.

[5] Cf. Whitehead, *Process and Reality*, pp. 121 ff.; 168-83.

[6] For this notion of causation cf. Whitehead, *Process and Reality*, esp. pp. 168-83.

[7] Whitehead, *Modes of Thought*, pp. 127-69.

[8] Cf. Whitehead, *Process and Reality*, pp. 18, 19, 22-26; 219-80, for an elaboration of the notion of prehension. For some of the paraphrasing here I am indebted to Lewis S. Ford, *The Lure of God* (Philadelphia: Fortress Press, 1978), esp. pp. 5ff.

CHAPTER IV: EMERGENCE

[1] For a lucid, but somewhat oversimplified statement of the traditional hierarchy of "levels" of reality cf. E. F. Schumacher, *A Guide for the Perplexed* (New York: Harper Colophon Books, 1978), pp. 15-38.

² With Paul Tillich, I prefer the term "dimension" to "level": "Under the dominance of the metaphor "level" the inorganic either swallows the organic (control) or the organic processes are interfered with by a strange "vitalist" force (revolt). . . " Systematic Theology Vol. III, p. 14. Although I shall occasionally resort to use of the term "level" for the sake of clarity, I shall usually employ the term "dimension" or "realm" in order to avoid the possible vitalistic interpretations that writers like Schumacher have fallen into by enslavement to the term "level." The term "dimension" allows for an interpenetration of the realms of nature more consistent with the "organismic" view we have been developing.

³ Francis H. C. Crick, Of Molecules and Men (Seattle: University of Washington Press, 1966), p. 10: note that Crick italicizes the word all.

⁴ J. D. Watson, The Molecular Biology of the Gene (New York: W. A. Benjamin, Inc., 1965), p. 67 (emphasis mine).

⁵ Gerald Feinberg, The Prometheus Project (Garden City, New York: Doubleday Anchor Books, 1969), p. 25. Crick's, Watson's and Feinberg's beliefs echo those of Jacques Loeb: "The ultimate aim of the physical sciences is the visualization of all phenomena in terms of groupings and displacements of ultimate particles, and since there is no discontinuity between the matter constituting the living and the non-living world, the goal of biology can be expressed in the same way." Quoted by John Hermann Randall, The Making of the Modern Mind (New York: Columbia University Press, 1976), p. 480.

⁶ Ernst Mayr, "Evolution," Scientific American Vol. 239, No. 3, (Sept. 1978), p. 50.

⁷ The following is to a great extent a restatement or paraphrasing of arguments given by Michael Polanyi in The Tacit Dimension (Garden

City, New York: Doubleday Anchor Books, 1967), esp. pp. 29-52; *Personal Knowledge* (New York: Harper Torch Books, 1964), esp. pp. 327-405; *Knowing and Being*, ed. by Marjorie Grene (Chicago: University of Chicago Press, 1969), pp. 225-39.

[8] The analogy of brick making is suggested by Polanyi, *The Tacit Dimension*, pp. 35 ff. However, I have taken considerable liberties with it here.

[9] Quoted by Sir Alistair Hardy, *The Living Stream* (New York: Harper and Row, 1965), pp. 265-66.

[10] Marjorie Grene "The Logic of Biology," in *The Logic of Personal Knowledge* (Glencoe Illinois: The Free Press, 1961), p. 199.

[11] The image of the flame is adapted from Polanyi, who in turn has borrowed it from W. Ostwald. Cf. *Personal Knowledge* pp. 384 ff.

[12] Polanyi, *Personal Knowledge*, p. 385.

[13] *Ibid.*, p. 384.

[14] Cf. Polanyi, *Personal Knowledge*, pp. 381-405.

[15] Hardy, p. 284.

[16] Barry Commoner, "In Defense of Biology," in *Man and Nature* ed. by Ronald Munson (New York: Dell Publishing Co., 1971), p. 44.

CHAPTER V: PURPOSE

[1] Again, in speaking of dimensions, I would prefer to employ the terms "surface" and "depth". However, since Polanyi employs the image of "level" together with the qualifications "higher" and "lower", I have, with reservations, adopted his terminology here.

2 Cf. Huston Smith, __Forgotten Truth__ (New York: Harper Colophon Books, 1977), pp. 1-177. For an elaboration of the principle that the lower is not adequate to the higher cf. Schumacher, __A Guide for the Perplexed.__

3 For Polanyi's use of the notions of "indwelling" and "reliance upon" cf. __The Tacit Dimension__ pp. 17-18; 30; 34ff.

4 Pierre Teilhard de Chardin, __The Future of Man__, trans. by Norman Denny (New York: Harper Colophon Books, 1964), p. 137.

5 __Ibid.__, pp. 133-134.

6 Ch. IV, n. 6.

7 Cf. Polanyi and Prosch, __Meaning__, p. 162.

8 Cf. Bernard Loomer, "Commentary on Theological Resources from the Biological Sciences," __Zygon__ I, 1966, p. 59.

9 Cf. Whitehead, __Adventures of Idea__, pp. 252-96. The following is a "loose" interpretation and application of Whitehead's theory of value.

10 I am indebted to David R. Griffin, __God, Power and Evil: A Process Theodicy__ (Philadelphia: Westminster Press, 1976), pp. 275-310, for much of my discussion of evil.

11 Whitehead, __Adventures of Ideas__, p. 265.

CHAPTER VI: PERISHING

1 For Whitehead's discussion of "perishing" see especially __Process and Reality__, pp. 340-41; 346-51.

2 Marcus Aurelius, __Meditations__, trans. George Long (Chicago: Henry Regnery Co., Gateway Edition, 1956), pp. 41, 44-45, 115.

3 *New English Bible*. (New York: Oxford University Press, 1970).

4 Quoted by Randall, p. 585.

5 Paul Tillich, *The Eternal Now* (New York: Charles Scribner's Sons, 1963), p. 33.

6 *Ibid.*, p. 34.

7 *Ibid.*

8 Cf. Whitehead, *Process and Reality*, p. 340: "This is the problem which gradually shapes itself as religion reaches its higher phases in civilized communities. The most general formulation of the religious problem is the question whether the process of the temporal world passes into the formation of other actualities, bound together in an order in which novelty does not mean loss."

9 Whitehead, *Science and the Modern World*, pp. 191-92.

10 See Whitehead's lecture "Immortality" in Paul A. Schillp, ed., *The Philosophy of Alfred North Whitehead* (New York: Tudor Publishing Co.), pp. 682-700.

11 Whitehead, *Process and Reality*, pp. 29, 60, 82 and *passim*.

12 *Ibid.*, p. 346.

13 *Ibid.*

14 John B. Cobb, Jr., *A Christian Natural Theology* (Philadelphia: Westminster Press, 1965), pp. 219-20.

15 Cf. Whitehead, "Immortality"; *Religion in the Making*, p. 149; *Process and Reality*, pp. 337-51.

16 Whitehead, Process and Reality, p. 344.

17 Ibid. pp. 67, 88, 108, 247, 347.

18 Ibid., p. 110.

19 Tillich, The Eternal Now, p. 35.

CHAPTER VII: ADVENTURE

1 Nikolai Berdyaev, Slavery and Freedom (New York: Charles Scribner's Sons, 1944), p. 87.

2 Ibid., P. 88.

3 Ibid., p. 89.

4 Ibid., p. 88.

5 Whitehead, Adventures of Ideas, pp. 258; 273-83.

6 Ibid., p. 274: "Advance or Decadence are the only choices offered mankind. The pure conservative is fighting against the essence of the universe."

7 Ibid. pp. 259-64; Whitehead, Science and the Modern World, p. 192; Process and Reality, p. 340.

8 Whitehead, "Mathematics and the Good," in Schillp, ed. p. 679.

9 Again Cf. Griffin, pp. 275-310.

10 Ibid.

11 Cf. Antony Flew, "Theology and Falsification: in Antony Flew and Alasdair MacIntyre, New Essays in Philosophical Theology (New York: The Macmillan Company, 1964) pp. 96-98; 106-108.

[12] Cf. especially Albert Camus' novel *The Plague* (New York: A. A. Knopf, 1948).

[13] Whitehead, *Process and Reality*, pp. 86, 88.

[14] *Ibid.* pp. 342 ff. Cf. John B. Cobb, Jr. and David Ray Griffin, *Process Theology: An Introductory Exposition* (Philadelphia: The Westminster Press, 1976) pp. 41-62.

[15] Birch, p. 76.

[16] Whitehead, *Process and Reality*, p. 342.

[17] *Ibid.*, p. 343.

[18] 2 Corinthians 12, 9: ". . . power comes to its full strength in weakness." *New English Bible*.

[19] Birch, p. 76.

[20] Whitehead, *Science and the Modern World*, p. 192: "The death of religion comes with the repression of the high hope of adventure."

[21] This is a major theme both in Whitehead's *Adventures of Ideas* and *Religion in the Making.*

[22] Whitehead, *Process and Reality*, p. 111.

[23] Whitehead, *Adventures of Ideas*, p. 276.

[24] Whitehead, *Process and Reality*, p. 105.

[25] Cobb and Griffin, p. 60.

[26] *Ibid.* p. 75.